The First 50 Years of the Department of Industrial and Operations Engineering at the University of Michigan: 1955–2005

The First 50 Years of the Department of Industrial and Operations Engineering at the University of Michigan: 1955–2005

Don B. Chaffin

Published in the United States of America by
Michigan Publishing
Manufactured in the United States of America

978-1-60785-367-1

An imprint of Michigan Publishing, Maize Books serves the publishing needs of the University of Michigan community by making high-quality scholarship widely available in print and online. It represents a new model for authors seeking to share their work within and beyond the academy, offering streamlined selection, production, and distribution processes. Maize Books is intended as a complement to more formal modes of publication in a wide range of disciplinary areas.

http://www.maizebooks.org

This book was produced using Pressbooks.com, and PDF rendering was done by PrinceXML.

Contents

	Preface	vii
	Acknowledgments	x
1.	The Early Days of the Department of Industrial and Operations Engineering	1
2.	The 1960s—The Beginning of a Contemporary Industrial Engineering Department at the University of Michigan	21
3.	The Maturing of the IOE Department in the 1970s	37
4.	The Department on the Move—the 1980s	51
5.	A Period of Growth and Change—1990–2005	69
6.	Building Centers and Programs	90
7.	The Center for Ergonomics	106
8.	Epilogue	121
	Bibliography	135
	Appendix	136

Preface

Why does someone commit to writing a history about anything? Hopefully, by doing so both the writer and readers will realize that good things don't simply happen; they result from the creative and hard work of many different people who shared a common vision and goals, often for many years. The Industrial and Operations Engineering Department has been a highly ranked leader in several different aspects of the field over its lifetime. Part of its success may lie in the fact that its faculty members have come from diverse fields, including mathematics, computer science, statistics, physiology, psychology, organizational psychology, mechanical engineering, and economics. In addition, many faculty members worked in or with a variety of industries before joining the department. These people appear to have used their diverse backgrounds in the belief that they could accomplish a larger impact by actively collaborating with others to solve a variety of major societal problems. This aspect of the culture within the department and its effect on the early success achieved by the faculty is one of the reasons for my writing this history.

A second motivation was to satisfy my curiosity about *why* certain developments and changes in the department took place over the past 60 years. In essence, I believe it is not enough to simply inform a reader about *when*, *what*, and *how* something important occurred, but it is more beneficial and interesting to discuss the circumstances and people that affected a certain event or change in an organization's behavior. This curiosity led me to read books about the history of technology and early scientific leaders and books by Anne and Jim Duderstadt that describe the history of the University of Michigan and the College of Engineering. I hope that by helping the reader understand more about why the department developed as it has, the reader will be better able to anticipate the challenges and opportunities that face the department in the near future.

Finally, on a personal note I wanted to understand at a much deeper level the organization that I have been part of for the past 50 years. Many people in this department have affected and shaped my career. They have been my teachers, mentors, friends, and colleagues. They have inspired and at times frustrated me, but always with good intentions. I believe that by writing about something one gains a much greater insight than by simply talking about it. So with that in mind I began this journey a couple of years ago, partly to inform others, but partly to simply understand my career at a much deeper level and the people who have influenced me.

I certainly hope that at some future time someone will see fit to update this book and that in the introduction they will say that most of what I have produced was helpful and truthful. That is all an author can expect when attempting to write such a history.

Don B. Chaffin

Acknowledgments

There are many people whom I wish to thank and acknowledge for their contributions to this text. In particular, there was the invaluable assistance of Elizabeth Fisher and Candy Ellis, who provided the minutes of past faculty meetings, scanned copies of IOE newsletters, provided past departmental reviews, and gathered the data and assembled the tables in the appendices that describe the expertise of and awards received by past and present faculty members and alumni. Elizabeth also deserves credit for her work in assembling the biographical information about many of the faculty members and alumni. I also wish to thank my good friend and colleague Steve Pollock, who contributed much of the text describing the history of operations research and reviewed and edited the text related to the financial engineering and ISDOS programs. To develop the history of the faculty's many different contributions to health care, I would be remiss to not thank Walton Hancock, a mentor, friend, and colleague who patiently responded to my many questions about this activity in the department. Bob Smith also took time out of his retirement to provide a description of the development and contributions that resulted from the work done in the department's Dynamic Systems Optimization laboratory. Much of the excellent history of the ISDOS program, as well as the description of its commercial development, was supplied by Hasan Sayani. Yavuz Bozer, Larry Seiford, and Jim Bean collaborated with Nancy Davis to provide a description of the roles IOE faculty members played in promoting and sustaining the Tauber Manufacturing Institute. Mark Daskin also deserves credit for providing critical comments and editorial advice regarding the epilogue. The wonderful photographs throughout the text were often obtained from the university's Millennium Project, developed and maintained by Anne and Jim Duderstadt, two people whose tireless efforts to capture and record the extraordinary history of the university and the College of Engineering have inspired the writing and publishing of this text. Rod Capps and Eyvind Claxton have also provided photographs from their collections. Lastly, I must thank Tom Armstrong, my colleague and friend in the Center for Ergonomics, for his assistance in describing the Rehabilitation Engineering Center Grant.

On a more personal note, I wish to thank my wife, Barbara, the joy of my life, for letting me know that retirement can be a good time to give back by remembering those who have meant so much in our lives. In many ways this book has been my way of saying thank you to her and the people in this institution who have enriched my life far more than I could ever have imagined or hoped for when I first visited in the spring of 1964. Soon after that visit I decided to undertake my PhD studies here. And the rest, as they say, has been a very good journey.

[1]

The Early Days of the Department of Industrial and Operations Engineering

The origins of industrial engineering in the United States arose because companies and government agencies needed to organize the workforce to be highly productive and competitive. The story of how and why people with different backgrounds came together to provide the means to accomplish this goal forms the basis for this chapter.

1.1 Early Pioneers of Industrial Engineering

Some have referred to industrial engineering as a "child of the Industrial Revolution." Indeed, around the turn of the 20th century, the transition in the United States and elsewhere from an agricultural society to a society that relied on mass production of goods and services of all kinds provided the need for a new type of engineering. With labor specialization and new manufacturing technologies came the need for better methods for planning and organizing how people and machines interacted to form an efficient manufacturing system. In engineering schools this began in the early shop courses that taught students how new types of physical and chemical processes could transform raw materials into finished parts and products. As companies took advantage of new manufacturing technologies, jobs for technically trained people became readily available in many different hardware-oriented disciplines. In the early part of the 20th century, engineering schools soon found themselves inundated with students eager to learn how to efficiently produce goods and services for a variety of rapidly expanding markets. For example, around the beginning of the 20th century there were several hundred small shops producing different cars, or what was then called a "horseless carriage," but by 1929, the US auto industry had consolidated many of these small shops into several very large corporations, including GM, Ford, Packard, Studebaker, Hudson, Dodge, and Chrysler. By carefully planning large production facilities, these companies were capable of producing more than 90 percent of the 32 million cars and trucks sold worldwide. And by 1930, this mass production capability resulted in there being one car for every five people living in the United States. The stories that follow are largely about how this type of transformation occurred and the role industrial engineering played.

The rapid industrialization, and its demand for engineers had a profound effect at the University of Michigan (UM). From 1880 to 1899 there were only four engineering degree programs at Michigan (civil, mining, mechanical, and electrical), and these granted a total of 530 undergraduate degrees and 30 graduate degrees over this19-year period. By 1950, the number of engineering programs at the UM had expanded to 22, and over a 12-year period from 1940 to 1952, these granted a total of 7,028 undergraduate engineering degrees and 2,176 graduate degrees.

As far as the practice of industrial engineering during the first few decades of the 20th century, it was mostly concerned with ways to improve worker productivity in various factories. This concern was facilitated by the writings of Fredrick Winslow Taylor and Frank and Lillian Gilbreth. In Taylor's book *The Principles of Scientific Management* (published in 1911), he described a process of selecting the most productive worker in a group, documenting exactly how this person performed a job, and carefully timing that person's movements. The combination of the careful documentation of worker motions along with the "time study," as it became known, were then used as the basis for specifying standardized methods when performing jobs and for selecting and training other workers on how to perform the job in an efficient manner. The resulting motion times were referred to as "standard times," which were provided to the managers and supervisors of a firm so they could better determine and manage labor costs, which often accounted for the majority of the total cost of a product. This was particularly true after Henry Ford shocked the industrial world in 1914 by offering to pay the best workers $5 per day (equivalent to about $120 today). This was twice the amount being paid to auto-assembly workers at the time, and it allowed Ford to hire the best mechanically skilled workers. This policy greatly reduced Ford's 300 percent annual worker turnover rate and, thus, was shown to be a cost-neutral policy.

In his many writings Taylor advocated that the results from a time study could be used to test and select well-trained and highly motivated workers to perform a job. Several cases were provided in his 1911 book that showed very large increases in productivity by using the results of time studies of "first class" workers in various companies. Of course, the results of a time study also became the basis for piece/part pay systems wherein additional bonus pay was provided to workers who exceed the production goal.

In 1912, Frank Gilbreth wrote a primer on how to use Taylor's principles. Gilbreth was an inventor, founder of a successful construction company, and later a management engineering consultant. He attributed a great deal of his early business success to Taylor's principles of scientific management. One of the major insights Gilbreth articulated was that because "standard times" were derived from adding up the sequence of motion times demonstrated by skilled workers, it was also possible to determine how to reduce the total time required to complete an existing job by reconfiguring the required motions in a manner that contained fewer wasted motions. Gilbreth described a simple example of this principle: in his construction company pallets of bricks were sorted and moved by low-paid, unskilled workers to be closer to the skilled bricklayers constructing a brick wall. The pallets of bricks were also positioned by adjusting the scaffold on which the bricklayer stood to a height that did not require them to do much stooping or reaching. This repositioning of bricks and worker not only reduced the time required of the skilled bricklayers to complete a wall, since less time was now required to walk and carry bricks, but also reduced the fatigue caused by repeatedly stooping and carrying the bricks.

The value of Taylor's scientific management principles became very well known and accepted after being used in a publically cited 1910 case before the Interstate Commerce Committee (ICC). In this action, the famous attorney Louis Brandeis solicited testimony from Taylor. This ICC forum allowed Taylor to publically describe his time and motion study methods and present his results from the time studies he performed for the railroads. The results swayed the ICC to deny the Eastern Railroad's request for a ticket price increase. The ICC stated that the projected labor costs were far higher than necessary due to management's inefficient use of workers.

Frederick Winslow Taylor (1900)

Together, Taylor and Gilbreth lectured between 1910 and 1915 at many different engineering and management conferences in the United States and abroad. Their common theme was that it was management's responsibility to use quantitative techniques (e.g., time and motion study) to document and standardize how best to organize jobs and to select and train the best workers. They and their followers quickly became known as "efficiency experts" or "industrial engineers." After the death of Frederick Taylor in 1915, Frank Gilbreth began to emphasize a slightly different philosophy than Taylor. He, with assistance from the Kodak Company, used the new medium of movies to film, sometimes at very high filming rates, skilled workers performing a variety of manual tasks. To do this, rather than time study workers in existing jobs, he established a laboratory where he could control the conditions in simulated work situations. In this sense he must be given credit as a pioneer for developing the first human motion capture laboratory in the United States. To acquire the time values that he needed he would place a clock with a large dial close to a worker and within the camera's field of view. In this way he could time their specific motions very precisely by reviewing the films. Gilbreth then categorized the motions into fundamental, or "elemental" motions (e.g., reach, grasp, move, position, walk, carry). To each of these he computed an average time based on his empirical laboratory studies. By summing the individual motion times required for a person to perform a particular job, he asserted that he could predict the mean total time it would take any skilled worker to perform a job. This motion time prediction method quickly became a powerful industrial engineering tool. By using the Gilbreth time prediction method a trained job analyst could simply describe a job on paper by listing Gilbreth's elemental motions. Since each of the listed elemental motions had an assigned time value, an analyst could now predict the total time for a person to perform a proposed new job. The Gilbreth motion-based time prediction methodology soon became part of the process used by engineers when designing new manufacturing operations. It provided industrial engineers with an estimate of the number and types of workers needed in a proposed factory, even before it was operating. It also provided a detailed standard description of the new jobs, which then improved the training of future workers.

Frank B. Gilbreth (1910)

After the much publicized Eastern Railroad case with the ICC in 1910, many industries took note of the writings and teachings of Taylor and Gilbreth. During the '20s and '30s these companies established motion analysis and time study departments staffed by industrial engineers. Improving the efficiency of worker performance, and thus reducing the cost of labor became a major goal of business executives during this era. Unfortunately, such a narrow focus had its limitations. In essence, workers were now being treated as a commodity in some factories. Supervisors would sometimes use time-study results to speed up a production line without regard to the mental and physical fatigue that it caused for some workers. Unionization of workers during the '30s in such plants was largely attributed to the unsafe and inhumane working conditions caused by the excessive profit motivation of managers. Unions, such as the United Auto Workers, which was formed in 1937, negotiated (sometimes after striking) a fair day's pay. They also required fatigue allowance times to be added to the standard time values. These fatigue allowances were based on how much workers slowed their motions during a day of work. Frank and Lillian Gilbreth's book *Fatigue Study: The Elimination of Humanity's Greatest Waste, a First Step in Motion Study* (1916) provided ample examples of how to determine if workers were becoming fatigued and how to reduce the amount of fatigue by redesigning the work conditions, such as providing stools to alleviate standing all day, arm rests on chairs, adjustable-height work tables, and adequate rest periods.

After the death of Frank Gilbreth in June 1924, his wife, Lillian Gilbreth, became a major spokesperson for those concerned with the use of motion and time studies in industry. While raising 12 children (their children Frank B. Gilbreth Jr. and Ernestine Gilbreth Carey chronicled the family's life in two books, *Cheaper by the Dozen* and *Belles on their Toes*, which were later made into successful movies), Lillian Gilbreth continued lecturing to industrial engineers from various companies. Lillian, who had earned a PhD in psychology from Brown University in 1915, emphasized in her many lectures and conference presentations that a rigorous, empirical approach was necessary to gather and analyze the complex human micro-motion data one needed to accurately predict the cost of labor and provide less fatiguing work conditions. She also developed and advocated the use of worker surveys to quantify worker attitudes and opinions, stating that valuable information could be gained from worker input to guide workplace improvements. In this sense she became an early spokesperson for more worker participation in decisions to improve work conditions. Her leadership became very important following the Great Depression of 1929. During the '30s and '40s she served on several different high-level government advisory boards in the Hoover and Roosevelt administrations, providing guidance on how to organize efficient and safe workplaces.

Lillian M. Gilbreth (1921)

For this pioneering work, in 1965 she was inducted into the National Academy of Engineering as its first woman member. In 1966, she received the Hoover Medal, an award given for "outstanding extra-career services by engineers to humanity." She has often been referred to as the "mother of ergonomics." Many different engineering awards have been named in her honor, including awards from the National Academy of Engineering, the Institute of Industrial Engineering, and the Society of Women Engineers, to name a few. She gave many lectures at various universities and was granted the title of professor of engineering at Purdue University in 1940. During her career she received 23 honorary degrees, the first of these from the University of Michigan in 1928. Lillian Gilbreth died in 1972 at the age of 93.

Based on the early adoption of time study and motion analysis methods by many different companies in the early part of the 20th century, the first industrial engineering undergraduate degree program was established in 1909 at Pennsylvania State University to further expand the availability of people to practice in this field. In addition to courses in accounting and shop planning and layout, the Penn State program offered a course on scientific management that used texts from Frederick Taylor and Frank and Lillian Gilbreth. How the Gilbreths influenced the development of industrial engineering at the University of Michigan is discussed later in this chapter.

1.2 Origins of Operations Research—The Use of Mathematical Modeling to Support Management Decision Making

This section provides a basis for understanding how operations research, that is, the use of mathematics to model and analyze a wide variety of systems and situations, has become an intellectual pillar, along with ergonomics and information systems, in the evolution of the IE Department.

In the late 1830s Charles Babbage investigated the cost of transporting and sorting mail, a project that in 1840 resulted in England's universal and highly efficient penny post. Much later, in 1908, two Swedish mathematicians—C. Palm and A.K. Erlang—established probability equations by which congestion in telephone systems could be represented. The resulting probability distributions of waiting times could then be used when designing the system.

In 1915, Ford W. Harris published a paper in *Factory: The Magazine of Management* titled "How Many Parts to Make at Once," in which he introduced what is now known as the "economic order quantity," a formula that is a cornerstone of modern inventory management. It allows a manager to balance the cost of reordering against the cost of manufacturing and maintaining a large inventory.

These are just three examples of the early use of mathematical modeling applied to operational problems. Improving, extending, and adapting these insights and results eventually led to whole fields of study and application: logistics, queuing theory (or, in England, "waiting line" theory), and production planning. Other similar precursors to contemporary approaches to analyzing operational problems can be found in the deep history of applied statistics and probability, game theory, graph theory, decision analysis, linear algebra, and so on, all of which have their antecedents in work performed by mathematicians in the 19th and early 20th centuries.

However, overshadowing these particular examples of the use of mathematics to address operational problems were the challenges to military decision makers in determining how to effectively operate in the world of 20th-century global warfare. This need for critical mathematical analyses of a wide variety of military problems eventually led to a discipline known by many names, including operations research, operations analysis, management science, and, more recently, analytics.

An early example of the use of mathematical modeling in the military is found in the 1916 book *Aircraft in Warfare, the Dawn of the Fourth Arm*, by English engineer F. W. Lanchester, which is a mathematical analysis of air-to-air combat that set the stage for a formal understanding of military conflict and its logistical components. (The resulting Lanchester equations, representing the evolution over time of power relationships among competing forces, are still used, not only in military analysis but also as a critical aspect of studying predator-prey situations). In 1917, Lord Tiverton showed the advantages of concentrating aerial bombs on a single target (a strategy leading to British and American forces using up to 1,000 planes in a single raid during World War II). Shortly after this, a study by the US Naval Consulting Board (headed by Thomas A. Edison) showed that the best way for surface ships to evade submarine attacks was to zigzag rather than to sail in a straight line.

In the unsettling years just before (and then throughout) World War II, the military establishments of the United Kingdom and their allies, including the United States, knew that winning the war depended on addressing operational issues. These included determining the best defensive and offensive actions against German U-boat attacks (which had been the cause of heavy losses of desperately needed supplies), finding the optimal ratio of escorts to merchant ships in cross-Atlantic convoys, determining locations for placing the newly invented radar towers and antiaircraft installations, developing the means to reduce the number of antiaircraft rounds needed to shoot down an enemy aircraft, and finding the most effective setting of the trigger depth of aerial-delivered depth charges.

Teams of mathematicians, statisticians, physicists, psychologists, and engineers in Great Britain, and then later in the United States, were charged with developing solutions for these kinds of operational problems. The involvement of first-rate scientists (including seven Nobel laureates) from a variety of fields created an environment within which their results had to be taken seriously.

The result was the development of a new military science by the end of World War II, called "operational research" in Great Britain and "operations research" (OR) in the United States. In the United States, military OR groups were first stationed at Princeton, Columbia, the MIT Radiation Laboratory, and other locations. Two important requirements for the success of the embryonic science emerged during this period: the use of multidisciplinary teams and the need for direct and prolonged field contact for analysts. Both the Navy's Operations Research Group (later called the Operations Evaluation Group, or OEG), which included around 70 analysts by the time the war ended in 1945, required their analysts to be connected to a fleet or wing "client" for six months every year, and thus were given direct access to decision makers at the highest levels. The Army Operations Research Group had similar requirements.

After the war, these teams provided the foundation for a postwar expansion of the tools and approaches of OR. Some evolved into organizations, such as the Rand Corporation (from the US Army's Project RAND) and the Center for Naval Analysis (from the OEG), that were formed to provide continuing scientific support for the military analysis of a wide variety of operational issues. Many others joined academic science, engineering, and mathematics departments at various universities.

George E. Kimball

Philip M. Morse (Photo courtesy of the MIT Museum)

Further academic interest in OR was evident in 1951, when Philip Morse (eventual chair of the MIT physics department, the first director of Brookhaven National Laboratory, and the founder and first director of the MIT Computation Center), along with George Kimball (a distinguished quantum chemist at Columbia University), published a newly unclassified book (originally written in 1945) titled *Methods of Operations Research*. This is believed to be the first comprehensive publication identified with the term "operations research." In this book they defined OR as "a scientific method of providing executive departments with a quantitative basis for decisions regarding the operations under their control."

Because of the authors' previous experiences, the various methods discussed in their seminal book were all directed toward a specific set of operations: those that involved challenges to the military arising from World War II. Other OR analysts, after leaving military or government service, found opportunities to apply their analytical and mathematical approaches to a host of nonmilitary problems. These new operations, instead of involving the sinking of submarines or determining bomb loads, were integral to production, logistics, manufacturing, extraction, scheduling, inventory control, distribution, competition, marketing, and so on. In the '50s the tools of OR began to be sought by a large number of firms, which motivated the hiring of well-regarded OR faculty members in the late '50s when the IE Department was formed at the University of Michigan, as discussed later in this chapter.

1.3 Early Development of Industrial Engineering at UM

The relationship between the Gilbreths and the University of Michigan began very early in the 20th century when UM professor of mechanical engineering Joseph A. Bursley met the Gilbreths at a 1910 conference on the topic of scientific management. After returning from this conference, Bursley began studying the topic in earnest, as he was convinced of its importance to future mechanical engineering students. He requested and was granted a two-year leave to consult with various companies, including the New England Butt Company, where the Gilbreths were instigating and studying the effectiveness of time and motion studies. He returned to UM and, in 1915, established a course in the Mechanical Engineering Department titled "Scientific Shop Management." During World War I this course was expanded to two courses, which were then required as part of the training of all Army ordinance officers. These two courses were so successful that the Army used them as a pattern for similar instruction at other universities.

Joseph A. Bursley (1924) – BS, Mechanical Engineering, University of Michigan (1899)

In 1921, Bursley became the dean of students at UM. In the same year Charles Burton Gordy was hired as an assistant professor of mechanical engineering for the purpose of continuing to teach courses on the topics of time and motion study.

After a series of committee reports and discussions, a new degree program in mechanical and industrial engineering was established in 1924. This new program required five years of study and 173 credit hours. Not surprisingly, eight years later only 14 students had graduated, probably because this bachelor of science degree required five years, whereas students in other engineering programs received a BS degree after only four years of study. As a result, the five-year degree program was replaced in 1934 with the awarding of a bachelor's degree in engineering (mechanical engineering) after four years and a master's degree in industrial engineering after the fifth year.

It should be clear from the preceding sections that in the early half of the 20th century industrial engineers were quite involved in organizing workers' required manual tasks in ways that allowed them to be highly productive. The need for efficient labor productivity was even more evident during World War II, when men and women needed to quickly learn new skills and perform very complex tasks in new war industries. The interest in designing products and work that were more compatible with human capabilities continued to increase after World War II. This was particularly true in Europe and Japan, where a large proportion of the population was impaired and required assistive devices and accommodations at work. The rapid development of new products during and after the war stimulated the formation of teams of engineers and life and behavioral scientists with the intent of better understanding a variety of human-hardware interface problems. In 1950 this resulted in a UK organization known as the Ergonomics Research Society, now the royally chartered Institute of Ergonomics and Human Factors. In the United States these multidisciplinary teams formed the Human Factors Society in 1957, which is now known as the Human Factors and Ergonomics Society. These organizations held annual conferences and workshops and facilitated the writing of human factors design guides and other documents and books, such as Ernest McCormick's 1957 textbook *Human Factors*.

But it wasn't only about developing better ways to utilize the labor force. In 1948 the American Institute of Industrial Engineers (in 1981 "American" was dropped from the name) was formed to promote the improvement and use of a variety of labor-planning tools as well as statistical and OR methods to improve productivity and the quality of manufacturing and service industries in the United States. By 1952, the increasing interest in military and nonmilitary OR led to the formation of the Operations Research Society of America (ORSA) as a means of advancing and diffusing developments in the field. This recognition of OR as a new intellectual discipline encouraged the establishment of educational programs devoted to developing and using the approaches and methods of OR. Depending on the nature of the institution, the academic homes for OR were usually offshoots of mathematics, statistics, business, or engineering departments (particularly industrial engineering), the latter being the eventual location of the University of Michigan's OR-related courses and research activities.By the mid-1960s, a wide variety of OR methodological approaches were developed, including linear programming (e.g., the simplex method for solving linear programming problems was developed in 1947 by George Dantzig, a decade after he received his MS in mathematics from UM), dynamic programming, queuing theory, game theory, network analysis, replacement and inventory policies, reliability analysis, machine maintenance, scheduling, Monte Carlo simulation, decision analysis, and stochastic modeling. These approaches were applied to a variety of problem areas in industry and in other arenas, including manufacturing, marketing, transportation, communications, public safety, construction, health care, medical systems, banking, military operations, entertainment, hospitality, financial engineering, logistics and supply chain management, and commercial aviation.

Charles Burton Gordy (1950) – BS and AM, University of Pennsylvania (1919)

1.4 Establishment of the UM Industrial Engineering Department

With the development of a sound statistical basis for time and motion study methods, combined with the newer operations research methodologies and, eventually, digital computing technologies, industrial engineering was beginning to be recognized as an important discipline in engineering schools and colleges in the United States. The effect at UM was that in 1946 the regents approved a degree designation of bachelor of science in engineering (industrial-mechanical); from then on, students could receive a BS degree in industrial engineering after four years of study. This degree option became very popular shortly after the end of World War II, especially with veterans, who saw it as a means to acquire jobs in the rapidly changing and highly competitive manufacturing sector.

In 1950, Gordy asked representatives from the Engineering Council for Professional Development (ECPD) to visit the program. They suggested several changes in the curriculum, namely a reduction in the number of accounting courses from the business school and an increase in courses in engineering economics, wage incentive and job evaluation methods, production control, plant layout, legal aspects of engineering, and statistics for engineers (the latter being taught by the Math Department). After UM made these changes, ECPD accredited the program. As a result, in 1951 the Board of Regents took the following action:

> On recommendation by the faculty of the College of Engineering, the name of the Department of Mechanical Engineering was changed to that of the Department of Mechanical and Industrial Engineering, with degrees of Bachelor of Science in Engineering (Mechanical Engineering) and of Bachelor of Science in Engineering (Industrial Engineering) to be awarded upon completion of the specified program.
>
> (November 1951 meeting from University of Michigan, Proceedings of the Board of Regents [1951-1954], page 227)

One year later, 98 students were enrolled in the new industrial engineering degree program within the Department of Mechanical and Industrial Engineering. Gordy, whose title was changed to professor of industrial engineering, became the program chair and was joined by assistant professors Quentin Vines, an expert in production planning; Wilbert Steffy, an expert in applied statistics and capital budgeting; and Edward Page, an expert in manufacturing processes and plant layout.

Quentin Vines – MS-ME ('53) University of Michigan

Wilbert Steffy – M.E.-ME&IE ('50) University of Illinois

Wilbert Steffy described the instructional program (*The Michigan Technic,* 1957) as follows:

> The program in Industrial Engineering in 1952 consisted of two options, "A and B." "Option A" could be defined as the field of planning, job specifications, job evaluation, time study, motion study, rate setting, incentive payment, plant layout, materials handling, production control, quality control, inventory control, employee rating, order procedures, packing and shipping, materials, salvage and waste reduction, and maintenance control. In explaining these functions of Industrial Engineering, Professor Gordy emphasized, on numerous occasions, that

the real industrial engineer is an enthusiast about the need for producing industry's goods and services at the lowest possible cost. He further subscribed to the philosophy that high wage rates had the best chance of existing in the plant in the industry using the most efficient methods in which labor strives to give the utmost instead of taking action to limit production. Also, that high wage rates can be paid when production per worker justifies them. "Option B" was intended to meet the need of those students whose interest lies in the field of manufacturing operations and methods. It includes the study of such processes as casting, forging, rolling, die casting, stamping, molding, machining, and the related functions of production planning, factory layout, processing, jig fixture and tool design, estimating for production, and inspection. Fabrication of material into finished parts was stressed in this option.

Edward L. Page – BS-IE&ME ('37) University of Michigan

It is interesting to note that in the early part of the '50s IE faculty research was mostly oriented to applied problem solving or consulting work in manufacturing companies. One exception was that the Methods-Time Measurement (MTM) Association, which supported a series of laboratory studies in the IE program. From 1953 to 1957, David Raphael worked with Gordy to perform research on the consistency of various standard time values within the MTM Association time prediction system, one of the most popular international systems for time prediction. His work led to a number of monographs concerning various aspects of predetermined motion times. This work was continued by Barbara Goodman when Raphael left in 1957. Walton Hancock, who was beginning as an assistant professor, then became the project director in 1960. In 1961, Hancock hired James A. Foulke, an electrical engineer, when Goodman left. The MTM Association project continued until 1972 and provided support for several PhD students under the direction of Hancock and Foulke. The emphasis also changed from determining time motion values to understanding the constraints on human productivity. By the end of the '60s, statistically determined human learning curves, error rates, operator selection, local muscle fatigue, aging effects, and complex decision times were being studied in the newly established Human Performance Research Laboratory, which had gained support from several companies as well as the MTM Association. In contrast to the MTM Association research, which was largely laboratory based, various companies funded students to perform field studies in their plants. It is worth noting in this context that in 1953 Robert Carson received UM's first PhD degree in industrial engineering with a dissertation titled "Consistency in Rating Method and Speed of Industrial Operations by a Group of Time-Study Men with Similar Training."

By 1955, it was clear to the faculty that students and companies had a strong interest in industrial engineering topics and principles. The American Institute of Industrial Engineering and the Operations Research Society of America were now publishing scholarly papers by faculty members from several universities. The original four UM faculty members (Gordy, Vines, Steffy, and Page), who were teaching in the production engineering program (Option A as described by Steffy) within the Mechanical and Industrial Engineering Department, had established a curriculum that was based on strong analytical problem-solving methods and an understanding of how workers could be expected to perform in various industries. In other words, the graduates from this program were now being provided with the type of knowledge required to solve important strategic and tactical problems in a variety of complex manufacturing and service operations. Based on this, at the October 1955 Regents meeting the following request to form a new Department of Industrial Engineering was approved as submitted by then dean George Granger Brown:

> The present Department of Mechanical and Industrial Engineering, including staff, other resources, and budget, is to be divided into two separate departments—Department of Mechanical Engineering and Department of Industrial Engineering. (From the October 1955 meeting, *University of Michigan, Proceedings of the Board of Regents [1954–1957]*, page 767)

Also, in 1955, Wyeth Allen was hired as professor of industrial engineering, He had expertise in organization management, had run an engineering consulting firm, and had served as president of a large manufacturing company. In the fall of 1956 the new Department of Industrial Engineering started offering courses with Allen as its first chairman. Several faculty members were added. In 1956, James Gage, an expert in engineering economics, and Richard Berkeley, who specialized in work measurement and labor planning. were hired. In 1957. Clyde Johnson, a highly respected management consultant was hired to broaden the curriculum to include organizational management and to develop student project courses for undergraduates.

Wyeth Allen – Hon Dr Eng (1953)

James Gage – UMBS, Mechanical Engineering, University of Wisconsin (1947)

Richard Berkeley – BS, Mechanical Engineering and Industrial Engineering, University of Michigan (1931)

Clyde Johnson – AB, University of Michigan (1931)

To develop the operations engineering area, in 1956 Robert Thrall, then a professor of mathematics at UM, joined the IE Department. In the previous decade Thrall had applied methods of mathematics and OR to military problems while consulting with the US Army Strategy and Tactics Analysis Group and the National Defense Research Committee. He also collaborated with John Nash on the potential application of game theory to military decision making. In 1962, he was instrumental in organizing the first US Army–wide Operations Research Symposium at the US Army Research Office, Durham, along with Dean Wilson, Merrill Flood, and Herbert Galliher (at that time the associate director of the MIT Operations Research Center and soon-to-be faculty member in the IE Department at UM).

Robert Thrall – PhD, Math, University of Illinois (1937)

At UM, further development in OR occurred a decade later when, in 1970, in collaboration with W. Allen Spivey (of the UM Business school), Thrall wrote *Linear Optimization*, one of the first textbooks devoted to linear programming. It should be noted that in 1969, Thrall was elected the 16th president of the Institute of Management Sciences (TIMS). During his term as president he was instrumental in founding and funding Center for Operations Research and Econometrics at the University of Louvain in Belgium, now recognized as a world center in OR. Thrall left UM in 1969 when he was asked to chair the new Department of Mathematical Sciences at Rice University, which had just been split off from the Department of Mathematics. While at Rice his research emphasis was the creation and development of data envelopment analysis.

Another important person in the OR area joined the IE department in 1959; Merrill M. Flood, who had received his PhD in mathematics from Princeton University. During World War II, Flood worked at the RAND Corporation, where he and Melvin Dresher conceived of and first analyzed the “Prisoner’s Dilemma” model of game theory, and the “Hitchcock Transportation Problem” (finding the minimum cost distribution of a product being supplied by several sources to several locations of use). He is also credited with naming the discipline of linear programming. Flood came to the University of Michigan with three titles: professor of industrial engineering, associate director of the Engineering Research Institute, and head of the Willow Run Laboratories (where he continued the work he started in the 1940s on combat surveillance systems). From 1959 through 1967 he was also professor of mathematical biology in the Department of Psychiatry in the medical school and senior research mathematician in the Mental Health Research Institute—foreshadowing the use of OR methods to study medical and health operational issues in which the department would become a leader. Flood was a founding member (and second president) of TIMS and, while at Michigan, was elected president of the Operations Research Society of America (ORSA). He was awarded the George E. Kimball medal in 1983 for his work with both societies and for his revolutionary contributions to OR and management science.

Merrill Flood – PhD, Math, Princeton University (1935)

Assisting Thrall and Flood in teaching OR during the late '50s was Richard C. Wilson, who was appointed an instructor in 1956 while completing his PhD in facility and production planning methods in the new IE department. In 1961 he was promoted to assistant professor and two years later to full professor. (See more about Richard Wilson in the next chapter.)

After IE became a separate department in 1955 and hired well-regarded and experienced faculty members, the effect on the curriculum was immediate. Teaching the Option B set of courses that emphasized manufacturing processes, described previously by Steffy, was largely left to the Mechanical Engineering Department. The Option A courses consequently became the IE department's primary teaching responsibility in the late '50s, with changes being made as new faculty members were added. By 1960, the BS degree in industrial engineering required the following set of courses, and more than 250 undergraduate students were enrolled in the program.

Though the focus of the department in its first few years was still on methods to improve manufacturing operations, expansion into other problem arenas had begun. For instance, Merrill Flood was providing new analytical solutions on how to optimize the routing of trucks, plan health care services, and perform product pricing in highly competitive markets. Clyde Johnson established an IE student project program to reduce the cost of hospital care. Johnson's pioneering hospital project course continues to this day, and over the years several faculty members have emphasized health care areas for their research in the department, as described in chapter 6.

IE Curriculum in 1960

Course	Credit Hours
First year:	
General English, math, and science courses	33
Remaining years:	
English, groups II and III	4
Econ. 153	3
Eng. Mech. 5: Statics and Stresses	4
Eng. Mech. 3: Dynamics	3
Chem.-Met. Eng. 18: Principles of Engineering Materials	3
Chem.-Met. Eng. 107: Metals and Alloys	3
Eng. Graphics 2: Descriptive Geometry	3
Elec. Eng. 5: D.C. and A.C. Apparatus and Circuits	4
Mech. Eng. 13: Thermodynamics and Heat Transfer	4
Mech. Eng. 14: Heat Power Laboratory	1
Mech. Eng. 31: Manufacturing Processes I	3
Mech. Eng. 33: Manufacturing Processes II	3
Mech. Eng. 83: Machine Design	3
Mech. Eng. 86: Machine Design II	3
Ind. Eng. 100: Industrial Management	3
Ind. Eng. 110: Plant Layout and Materials Handling	3
Ind. Eng. 120: Work Measurement	3
Ind. Eng. 130: Wage Incentives and Job Evaluation	2
Ind. Eng. 135: Management Control	3
Ind. Eng. 140: Production Control	2
Ind. Eng. 150: Engineering Economy	2
Ind. Eng. 160: Operations Research	3
Ind. Eng. 165: Data Processing	3
Bus. Ad.: Accounting 100	3
Bus. Ad.: General 154: Industrial Cost Accounting	3
Math. 161, 162, Statistical Methods for Engineers	6
Nontechnical electives	7
Technical electives	8
Total	128 hours

1.5 Changes in the College of Engineering after World War II—Effects on IE

One change that affected the new department in the 1950s was a major philosophical shift that had begun after World War II within the College of Engineering leadership. Up until this time the major focus was almost exclusively on undergraduate education, mainly because that is what tuition and the state and federal government were funding. In the latter regard, the Servicemen's Readjustment Act of 1944, which became known as the GI Bill of Rights, provided tuition payments for veterans and assistance with living expenses and books while studying toward a degree. During 1947, veterans accounted for almost half of all college enrollees, and by July 1956 about 7.8 million veterans of the 16 million that had fought in World War II had taken advantage of this program, thus making a college education available to those other than the rich in the United States.

Indeed, undergraduate enrollment in the college went from 2,155 in 1940 to 3,273 by 1970. With the hiring of additional faculty members in the college, many of whom had PhD degrees, by the early '50s there was a growing emphasis on graduate education and research as well. The result was that graduate enrollment increased from 258 students in 1940 to 948 in 1970. As the new IE Department hired additional faculty members in the late '50s and better research facilities were provided, the IE graduate programs began to develop. By 1957, an MS-IE degree was offered, though it did not have specific course requirements. In fact, the *1960 Engineering Bulletin* stated: "A candidate for this degree must have completed satisfactorily the undergraduate industrial engineering degree or its equivalent, and must complete in residence a minimum of thirty hours of recognized graduate work approved by the adviser. The course selections necessary for this degree are rather flexible, but it is expected that approximately twelve hours of course study will be in the industrial engineering area."

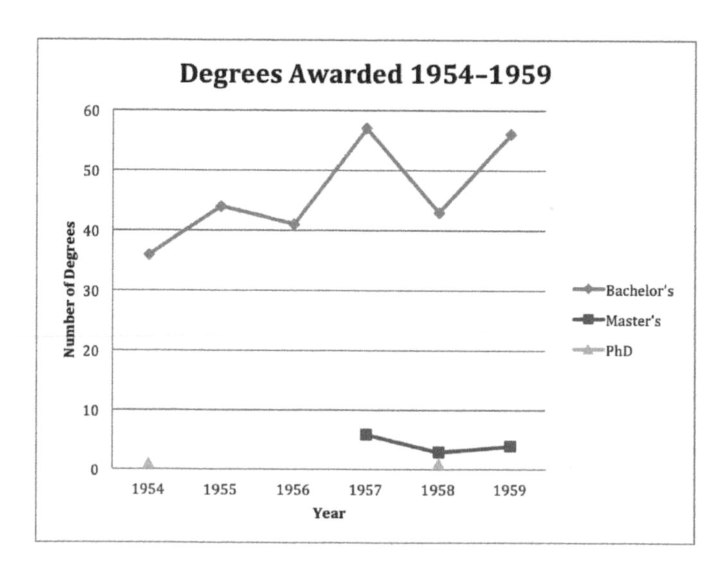

The IE program graduation profile in the graph displays this trend, with the early emphasis on providing undergraduate degrees followed by the initiation of the graduate program when the new department began in 1955.

The emerging need to provide a strong graduate education in the late 1950s, which soon became a formal second function of the college, had a very important goal: to establish a first-class research university. Doing so was contingent, however, on the development of sponsored research because Michigan legislators were much more interested in providing well-educated engineers with BS degrees. For instance, federal and private funding of the college's research was only $215,700 for the 1940–1941 academic year (equivalent to about $3.6 million today). But with the growing emphasis on research, and with the hiring of new faculty who had PhD degrees in the 1960s, there was the interest and capability to develop more sponsored research. Fortunately, following World War II many companies needed the results of engineering research to guide the development and use of new technologies and methods of conducting business. The federal government had also become accustomed to supporting university research during World War II, and this continued with the establishment of the National Science Foundation in May 1950. As a result, by the 1965–1966 academic year, sponsored research funding from all sources had reached about $10 million in the College (equivalent to $74 million today), which was more than double the state allocation for the entire general operating fund for the College. In essence, the College rather quickly became dependent on the faculty's ability to attract major sponsored research funding. As will be discussed, the IOE faculty responded in kind to this policy.

Another College-wide change during the 1950s was the beginning of the North Campus development. The resulting new building space enabled the College to integrate science and research with graduate engineering education by establishing a variety of laboratories. The Cooley Laboratory, dedicated on October 24, 1953, was the first building on North Campus, and it provided space for some important developments in electronics, much of which had been done previously as classified, military-oriented research at the UM Willow Run Laboratory during World War II. Over the next few years, several other engineering buildings were completed, including the Ford Nuclear Reactor in the new Phoenix Laboratory in 1955. This facilitated a new educational program in nuclear engineering, one of the first in the United States. Other buildings quickly followed, including the Automotive Engineering Laboratory in 1957, the Aeronautical Engineering Laboratory and Wind Tunnel in 1957, and the Fluids Laboratory (later renamed the G. G. Brown Laboratory Building in honor of the dean who initiated the building construction on North Campus), which followed a 1956 plan that had been proposed by Dean Brown and approved by the regents. In this plan, all the functions of the College were to move to North Campus in phases. Unfortunately, it was not until the 1980s that such a move took place.

The delay in the construction of the additional buildings necessary to provide the space for all of engineering to be located on North Campus had a major effect on the IE program. Fortunately, the new IE program initiated in the mid-1950s required only a modest amount of dedicated laboratory space in West Engineering, mainly for the research sponsored by the MTM Association. Almost all of the other department space at that time was used for offices and classrooms. This was to change in the '60s with the growing interest in solving a variety of human performance problems that were plaguing companies at the time.

West Engineering and the "Arch" were home to the IE Department from 1955 to 1983.

1.6 Synopsis of 1950s—IE Contributions and Major Events

- 1955: IE department separates from the Mechanical and Industrial Engineering Department. Four faculty members (Gordy, Vines, Steffy, and Page) from the Mechanical and Industrial Engineering Department become the founding members, with Richard Wilson as a lecturer.
- 1956: The undergraduate curriculum is established and the first IE courses are taught. Six new faculty are added to the department (Allen, Gage, Berkeley, Thrall, Flood, and Johnson).
- 1957: Johnson establishes a senior project course in hospital systems, IE undergraduate enrollment is now 260 students, and several OR courses are added to the curriculum at the undergraduate and graduate levels.
- 1958: Graduate-level courses are established in manufacturing process planning, plant layout, optimization methods in IE, hospital systems, and quality control and inspection. These courses enroll about 20 master's-level students in Ann Arbor. The part-time evening master's program enrolls 18 students in Flint. The first IE Department PhD student graduates (Fritz Harris), supervised by Allen.

[2]

The 1960s—The Beginning of a Contemporary Industrial Engineering Department at the University of Michigan

2.1 General Trends in IE Education and Research—A Period of Growth

By 1960, the newly formed Industrial Engineering (IE) Department was certainly part of the larger trends in the College of Engineering. Not only had the undergraduate program grown during the late '50s but in the '60s the graduate programs became very important. This can be seen in the graph depicting the annual IE graduation rates during the '60s. Some of the large increase in the number of master's degrees awarded in the late '60s was due to the fact some of the faculty were teaching night classes at the University of Michigan Flint Campus and at a community college in Saginaw, Michigan. About 20 MS students graduated annually during the late '60s and early '70s through this IE extension program. It was discontinued in 1982 due to low enrollment and lack of interest in teaching the off-campus courses in the evening. Some would say this was the beginning of the department's withdrawal from actively promoting and supporting continuing education for professional engineers.

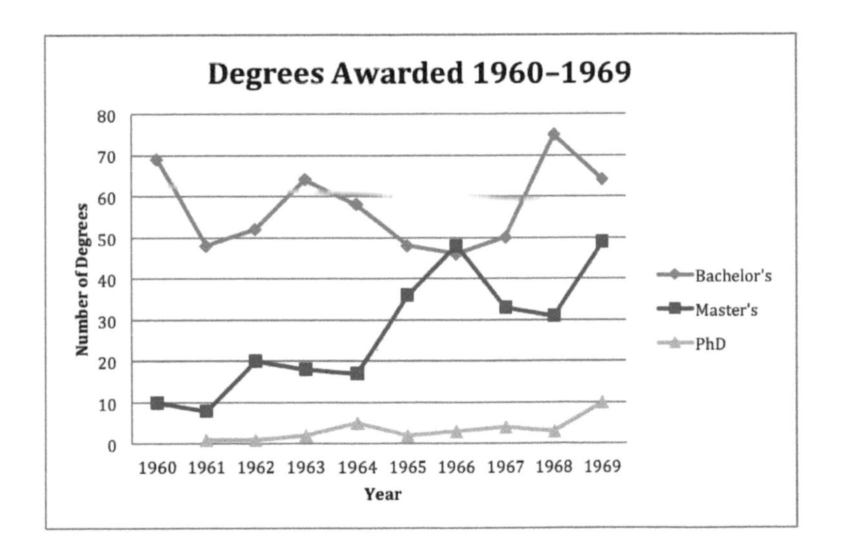

Providing the means to expand the PhD program in IE required funding for research, and several IE faculty members began to aggressively seek funding. In 1962, the total sponsored research funding in the department was about $25,000, but by the end of the '60s more than $700,000 was acquired annually from various companies and federal agencies (the equivalent of about $4.2 million today). This funding allowed the more research-oriented faculty to "buy out" from teaching a course or two during the academic school year and supported additional PhD students, a trend that continued through the '70s and '80s. Such fund-raising work was further necessitated by the diminishing university support for the College during the '70s, which will be discussed later.

2.2 The Need for a Broad IE Discipline

During the early '60s many technological changes were occurring in the world. The race to understand outer space was in full swing, with satellites flying overhead and the launching of spacecraft that flew astronauts to the moon and back as part of the NASA Apollo Program. The federal government continued to develop very sophisticated military equipment, including high-altitude spy planes to help determine what the rest of the industrialized world was doing. Most importantly, programmable digital computers were rapidly proliferating in universities and companies. These allowed complex computational problems to be solved in hours that had taken days, or worse, could not be solved at all without the use of these new computers. It clearly was an exciting time for science and technology. In many cases, new research results were completely changing the work, play, and lifestyles of most people living in the industrialized world, and IE education in the United States was rapidly expanding, with 48 degree-granting departments in operation by the end of the '60s.

2.3 The Importance of the Computing Environment in the Department

Prior to the widespread use of digital computers in academia, special digital processors were built to solve specific types of engineering problems. In 1952, the Michigan Digital Automatic Computer (MIDAC), built by the University of Michigan (UM) Willow Run Research Center, was the first digital computer at the university. Several MIDACs were built and used for solving linear-equation problems in the 1950s, including transportation problem algorithms developed by Professor Merrill Flood and facility layout algorithms of interest to Professor Richard Wilson.

Richard Wilson using a UM-built MIDAC in 1961 in West Engineering.

In the '50s and early '60s, more flexible, programmable digital computers were quickly spreading to businesses and universities. Unfortunately, programming such machines was still time consuming and often required instructions and data to be input on hundreds of key-punch cards. These cards were then submitted to a computer center that had the machines to read the hole patterns in the cards and run the program on a mainframe computer. The results would be returned to the programmer as a set of printouts.

In 1952, the university acquired a card-based computer, the IBM Card Programmable Computer. In 1956, the first generally available computer, an IBM 650 was acquired and set up on Central Campus in a new computer center that would serve all faculty and students. This was followed by a succession of larger and more powerful computers over the next 10 years, but all required students to bring their punched cards to the center and wait, sometimes hours, for the staff to run their programs. As computer interest and programming capabilities exploded on campus, two developments relevant to IE took place. First, in 1965 a time-sharing capability was provided to users with remotely located teletype-style input/output terminals being connected to the mainframe computer, an IBM 360. A few of these terminals were located in West Engineering, so IE students and faculty could now type in their instructions and data rather than trudging to the computer center with a box full of punch cards. Unfortunately, the fees to use the remote terminals were about three times higher than that for running a program in a batch mode at the computer center.

The second development was that computer companies like Digital Equipment Corporation and Hewlett-Packard began to build small, stand-alone, scientific computers. Such a machine was acquired by the IE department in 1963 and used by Wilson and others to develop algorithms for solving transportation, production scheduling, plant layout, and other problems. By 1969 these mini computers were being used to monitor experiments online, and some were interfaced with the mainframe computer for later data processing. For example, an HP2115A mini computer was installed in the G. G. Brown Laboratory Building and used by the IE human performance laboratory and the nuclear engineering, bioengineering, and mechanical engineering departments. In addition, James Foulke (a very creative electrical engineer) worked with Professor Walton Hancock to create a special, programmable, digital data calculator that was configured to collect and analyze human motion and time data and was interfaced with mini computers for further data analysis.

Richard Wilson using the department's new Litton General Purpose LGP30 Digital Computer (with a drum memory) in 1963 to run plant layout and transportation models in West Engineering.

Many years later, in the mid-1980s, the College established a sophisticated network known as the Computer Aided Engineering Network (CAEN) to support various types of computers, including the new personal computers. Now all faculty, staff, and students had access to computers to solve many different types of engineering problems. This was further aided by the College administration's negotiation with Apple Corporation to provide up to 60 percent discounts for the Macintosh personal computers. From then on most large engineering problem solving was accomplished with the aid of analytical models and associated algorithms that could be run on individual and network-supported personal computers.

The HP2115A computer used to monitor real-time experiments in 1969 on North Campus.

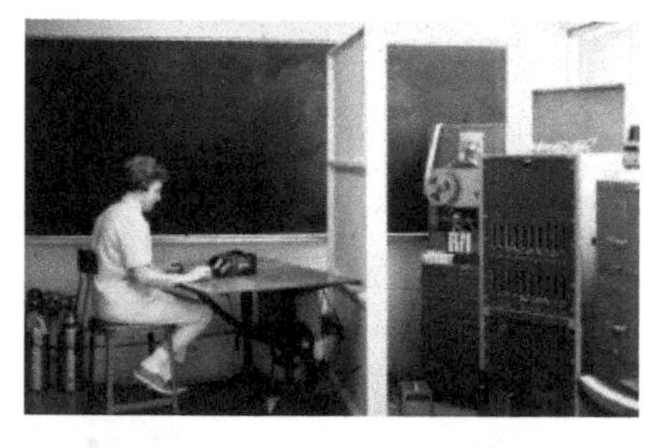

The data calculator used for MTM Association research in 1969 in the G. G. Brown Laboratory Building on North Campus.

2.4 Changes in IE—More Research, More Graduate Students

The rapid changes in manufacturing, transportation, construction, health care, and other industries in the 1960s meant engineering students had to be cognizant not just of a specific type of new technology but also how that technology functioned in unison with other existing technologies commonly used in a particular industry. Managers were asking questions such as the following: (1) What costs would be incurred when adopting a new technology? (2) How will a proposed technology be used to enhance the manufacturing, operation, and maintenance of new products and services? and (3) How could a variety of people benefit from using a new technology? Answers to these and other related questions led to the hiring of additional faculty members in the department. These faculty members expanded the breadth of IE courses, which then provided a broad technical education at the undergraduate level and a certain amount of specialization at the graduate level, the latter relying on a defined set of optional courses or "tracks" in various subspecialty areas in which the faculty had particular expertise.

2.5 Faculty Changes

Teaching courses to address the types of questions posed earlier at both the undergraduate and graduate level required the hiring of additional faculty members. These new faculty members were highly committed to conducting their own research while teaching courses and advising PhD students. The department had six full-time, tenure-track professors by the end of the 1950s, 10 by 1962, and 18 by 1969. Of the faculty members who joined the department during the 1960s, the following had a major and lasting effect.

Walton M. Hancock – PhD,
Johns Hopkins (1954)

Walton Hancock was hired in 1960. His research stressed the understanding of how people perform and learn complex perceptual-motor tasks with emphasis on statistical modeling of human performance in manufacturing systems. He soon established a human performance laboratory that broadened time and motion study to include statistical models of human performance variations caused by different types of training, age, and fatigue factors. One of his many original contributions was the learning curve, a statistical method for forecasting how quickly workers could be expected to learn new skills. In 1963 he was promoted to department chair, a position he held until 1968. During the 1970s he used his analytical modeling methods to improve patient scheduling and hospital staffing processes and held a joint appointment in the Department of Hospital Administration within the School of Public Health. Later, Hancock served as associate dean of the College of Engineering and director of the Center for Research on Integrated Manufacturing from 1985 to 1989. (See Hancock's memoir at the UM Faculty History Project: www.um2017.org/faculty-history/faculty/walton-m-hancock.)

Richard C. Wilson – PhD,
University of Michigan (1961)

Richard Wilson, who had begun his academic career at UM as an instructor in 1956, was promoted to professor in 1961. As an instructor in the late '50s he had taught the use of digital computers for solving various IE problems. He was soon acknowledged for his expertise in providing useful solutions to complex analytical models related to facility layout, production planning, and production scheduling. In the mid-1960s he established a production simulation laboratory for teaching and research related to facility layout and production scheduling. He later served as associate dean in the College of Engineering from 1968 to 1972 and chair of the department from 1973 to 1977. In 1978, he established a collaborative education and research program in manufacturing with mechanical engineering and computer science that later became the Center for Research on Integrated Manufacturing.

Herbert Galliher – PhD, Yale (1952)

Herbert Galliher was hired in 1963 from MIT, where he had served as the associate director of the pioneering Operations Research Center for seven years. When he came to Michigan, Galliher taught and developed courses in inventory and production analysis, stochastic processes, and mathematical modeling. As one of the founders of the field of operations research, his research and teaching covered a range of applications. He developed and named the "stuttering Poisson" distribution to describe demand processes in supply situations. He was one of the first to use queuing theory to analyze landing congestion of aircraft and to use stochastic models to represent disease processes, leading to an early and influential body of literature on the management of cervical cancer, arteriosclerosis, and breast cancer. The results of his project, *Aircraft Engines: Demand Forecasting and Inventory Redistribution*, as well as the development of a military depot system simulator revolutionized the forecasting of the demand for spare engines and associated logistical decisions. Galliher was the founding editor of *International Abstracts in Operations Research* and associate editor of *Operations Research* and the *SIMA Journal on Applied Mathematics*. He was also a well-respected consultant to a variety of corporations and government agencies. (See Galliher's obituary in the December 11, 2000, issue of the *University Record*: www.ur.umich.edu/0001/Dec11_00/22.htm.)

In 1965, **Bertram Herzog** transferred to the IE Department from the Department of Engineering Mechanics, where he had become an expert in the use of computer graphics for the design of complex structures. In 1968, shortly after joining the IE faculty, he became the first director of the new Michigan Educational Research Information Triad (MERIT) computer network, one of the first operational computing networks in the country. Under his direction, MERIT connected the mainframe computers at the three large public universities in Michigan in 1972. This enabled faster computational processing than had been possible and the rapid sharing of large data files. By the mid-1980s, the MERIT staff started managing the connectivity of various computers around the country as part of the NSFNET, the precursor of today's Internet. (See Herzog's obituary on the Electrical Engineering and Computer Science Department website: http://www.eecs.umich.edu/eecs/about/articles/2008/Herzog.html.)

Bertram Herzog – PhD, Engineering Mechanics, University of Michigan (1961)

Don Chaffin – PhD, University of Michigan (1967)

Don Chaffin, who had been an instructor in the department while studying for his PhD (1964–1967), returned in 1969, after serving as a faculty member at the University of Kansas Medical Center. Working with Hancock, he enlarged the scope of the Human Performance Laboratory to include the study and prevention of working conditions that raised the risk of physical fatigue and injury to a worker's musculoskeletal system. From 1977 to 1981, he served as department chair. In 1982, he assumed directorship of the Center for Ergonomics (discussed in chapter 6), remaining in this position until 1998. From 1998 until his retirement in 2007 he directed the Human Motion Simulation Laboratory within the Center for Ergonomics. Chaffin is a fellow in eight scientific and professional societies, was elected to the National Academy of Engineering in 1994, and received the National Engineering Award from the Association of Combined Engineering Societies in 2008. (See Chaffin's personal page on the UM website: www-personal.umich.edu/~dchaffin/.)

Katta Murty joined the department in 1968 from the University of California, Berkeley, replacing Robert Thrall, who had taken a position as chair of the Applied Mathematics Department at Clemson University. Like Thrall, Murty had a deep understanding of and commitment to furthering the development of linear and nonlinear optimization methods and algorithms. His eight textbooks have been used as the basis for teaching optimization methods and applications to students around the world. (See his personal page on the UM website: www-personal.umich.edu/~murty/.)

Katta Murty – PhD, University of California–Berkeley (1968)

Daniel Teichroew – PhD, University of North Carolina (1953)

Daniel Teichroew joined the department in 1968 from Case Western Reserve University. Before that he was a professor at Stanford University. By the time he came to UM, he had already published six books on management science with an emphasis on computer algorithms and languages. Teichroew was hired as the department chair, a position he held until 1973. During this period he continued his pioneering development of a Problem

Statement Language (PSL), which was the intellectual core activity within the Information System Design and Optimization System (ISDOS) consortium that he had begun building while at Case Western Reserve. The resulting ISDOS software and technical reports provided an efficient way for a person to use a computer to completely specify the information technology (IT) requirements for a proposed engineering or management system. In essence, the ISDOS reports provided a set of coding rules that would guide the formal statement of a technical or managerial problem in such a manner that a computer algorithm, known as a Problem Statement Analyzer (PSA) could evaluate the PSL problem statements and thus determine if the IT requirements were met. (For more on ISDOS, see chapter 6. See also Teichroew's memoir at the UM Faculty History Project: http://um2017.org/faculty-history/faculty/daniel-teichroew/memoir.)

Ralph Disney – DEng, Johns Hopkins (1964)

Ralph Disney came to the university in 1962 as a visiting professor from the University of Buffalo (now the State University of New York at Buffalo). He rose quickly to full professor at UM after completing his DEng degree in 1964 from the Johns Hopkins University. His adviser at Johns Hopkins was a well-known expert in applied probabilism, Bruce Clarke, who soon after joined the UM Mathematics Department. Disney and Clarke continued their collaboration at UM and authored two textbooks on applied probability and stochastic processes in the early '70s. His research contributions were in queuing theory (particularly queuing networks such as those found in manufacturing, transportation, and communications) and the mathematics of stochastic processes. He produced more than 70 research articles and was a founder of the Applied Probability College of Operations Research Society of America. In 1977, Disney accepted an invitation to be the Charles O. Gordon Professor of Industrial Engineering at the Virginia Polytechnic Institute and State University in Blacksburg. Disney was recognized by the Institute of Industrial Engineers (IIE) with the David Baker Award and the Albert Holzman Award for teaching, research, and other contributions to his field as well as the Frank and Lillian Gilbreth Award, the highest honor bestowed on an industrial engineer. In 1999, Disney was elected into the National Academy of Engineering. (See Disney's obituary from the November 15, 2014, *Washington Post* here: http://www.legacy.com/obituaries/washingtonpost/obituary-print.aspx?n=ralph-l-disney&pid=173179805.)

Seth Bonder – PhD, Ohio State University (1965)

Seth Bonder joined the department in 1965, after receiving his PhD in industrial engineering from the Ohio State University. From 1952 to 1956, he served as a pilot for the US Air Force in the Korean War. From this experience and his education, Bonder became a major leader in the use of operations research methods to solve very large and complex military problems of all sorts. In 1972, he left the department to establish and become chief executive officer of one of the most successful military OR consulting companies in the United States. The company, Vector Research Incorporated (VRI) grew to more than 400 employees, including several PhD graduates from the department. His work garnered him many national awards, and he was elected into the National Academy of Engineering in 2000. Before his death in 2011, he shifted some of his considerable talent to directing VRI programs in health care. He also endowed a scholarship program in the IOE Department. (See the INFORMS website for a tribute to Bonder written by Stephen Pollock: https://www.informs.org/ORMS-Today/Public-Articles/December-Volume-38-Number-6/In-Memoriam-Seth-Bonder-1932-2011.)

Stephen Pollock joined the department in 1969. Before that he was on the faculty of the US Naval Postgraduate School. His PhD from MIT provided him with a deep understanding and background in a variety of empirical and probabilistic-based operations research methodologies. He used these insights to collaborate with many other faculty members in solving problems related to (1) public systems, for example, political redistricting, nuclear arms, and collusion monitoring; (2) medical systems, for example, diagnostic methods, staffing levels, and early infection detection; (3) manufacturing, for example, supply routing for ship building, motor vehicle emission testing, and spot welding classification; and (4) sporting events, for example, swimmer event assignments and golf handicapping. From 1981 to 1990 he served as department chair. His stochastic operations modeling work in a variety of complex problems gained him a number of national awards, and he was elected to the National Academy of Engineering. (See his page at the College of Engineering: http://pollock.engin.umich.edu/.)

Stephen Pollock – PhD, MIT (1964)

Richard Jelinek – PhD, University of Michigan (1965)

Richard Jelinek joined the department in 1965 after completing his PhD on health care planning methods. Working with Clyde Johnson, he developed several well-funded research studies that demonstrated how a variety of industrial engineering methods could be used to improve the quality and cost-effectiveness of local hospitals. In 1970, he cofounded the Medicus Company to apply the results of these studies in a wide variety of hospitals, and in 1971, he joined the faculty in the UM School of Public Health, Bureau of Hospital Administration. Jelinek later became chief executive officer of AmeriChoice Corporation, a national health care consulting company. (See his executive profile on the BloombergBusiness website: http://www.bloomberg.com/research/stocks/private/person.asp?personId=228637&privcapId=24792&previousCapId=33254122&previousTitle=RedBrick%20Health%20Corporation.)

Among the other faculty members who joined the department for short periods during the 1950s and 1960s were Edgar Sibley (1966–1972), who was hired as an associate professor and worked with Daniel Teichroew, and Dean H. Wilson (1961–1968), an associate professor hired to assist in teaching the use of digital computers to control manufacturing systems and to develop and teach the use of Monte Carlo simulation methods. Assistant professors included Fred Black (1955–1958), Richard Evans (1958–1963), Hugh Bradley (1965–1968), Stephen Kimbleton (1968–1973), and Richard Baum (1969–1976). Richard Pew, an assistant professor in psychology, had a joint appointment in the department from 1966 to 1972.

On average there were 17 faculty members in the department during the 1960s, and these provided a certain amount of structure and stability in the curriculum; prerequisite courses were offered more often than before, and more advanced courses were provided to support the growing graduate program. But toward the late 1960s, the end of the Apollo space program, the contentious war in Vietnam, and greater demands for racial equality in various organizations (including higher education) led society to lose interest in technology, and the University of Michigan executives began a process of shifting financial support away from the College of Engineering and toward the liberal arts, social, and health-related programs. This defunding of the College, which continued throughout the '70s, meant restrictions on hiring new faculty, limited salary increases, and difficulties in achieving promotion to tenured positions for the existing new faculty members, especially if the candidate did not have a large amount of sponsored research funding to support his work and students.

In regards to this latter situation, even though the number of full-time faculty members had increased from nine to 18 in the '60s, the number of funded positions on the general operating budget supported by the state and tuition funds only increased by one during the same period. In other words, there was an enormous burden on the faculty to acquire grant and contract funds to support themselves and their graduate students. In fact, the college administration expected all tenured faculty members to provide at least 20 percent of their academic-year salary from external funds. The IE faculty for the most part responded by ramping up the externally funded research from about $25,000 in 1962 to more than $700,000 in 1970 (the equivalent of about $4.2 million in 2014). This relatively large amount of sponsored research in the department supported at least 43 PhD students by the end of the '60s and allowed the faculty to grow to 19 members at the beginning of 1970 without placing a major burden on the diminished college general fund budget.

2.6 Beginning to Move to North Campus

The increasing sponsored research funding throughout the '60s also allowed some of the empirically oriented IE faculty members to establish a few laboratories for instruction and research purposes in the G. G. Brown Laboratory Building on North Campus. About 2,000 square feet was dedicated for these IE laboratories and offices by the mid-1960s. This relatively small space accommodated about 15 PhD students and five faculty members for the purpose of physically simulating and studying production and assembly operations as well as conducting a variety of human performance experiments.

Unfortunately, the very tight college budgets provided by the state starting in the mid-1960s and continuing through the 1970s resulted in almost no funds being available to construct the buildings needed to move more of the College operations to North Campus. As a result, faculty and students in many departments, including IE, had to commute daily between the instructional spaces in West and East Engineering Buildings on Central Campus and their research offices and laboratories on North Campus. This situation was finally resolved when the entire IE faculty moved to North Campus in 1983 (see chapter 4).

The west end of the G. G. Brown Laboratory Building in 1964 is shown in the foreground; this is where IE offices and laboratories were located on the North Campus.

2.7 1970 IE Curriculum

By the end of the 1960s, the undergraduate IE curriculum continued to emphasize a balance between traditional, hardware-oriented engineering technologies that dealt with materials, structures, chemicals, electricity, and electronics and the more analytical and statistical methods needed to solve multifactor problems in highly complex systems composed of people operating and servicing hardware and software systems of all kinds. Candidates for the degree Bachelor of Science in Engineering (Industrial Engineering) had to complete the following program:

IE Curriculum in 1970

Course	Credit Hours
Subjects required by all programs	(56 hours)
Mathematics 115, 116, and 117, 215, and 216	16
English 101 and 102: Great Books I and II	6
Eng. 101: Graphics; Eng. 102: Computing	4
Chemistry 103 or 104	4
Phys. 140 with Lab. 141 and 240 with Lab. 241	8
Literature and Rhetoric	6
Humanities and Social Science	12
Statistics (6 hours)	
Stat 31: Elem. of Prob. and Math. Stat. I	3
Stat. 311: Elem of Prob. and Math. Stat II	3
Related technical subjects (15 hours)	
Eng. Mech. 211: Intro. to Solid Mech.	4
Mat.-Met. Eng. 250: Prin. of Eng. Materials	3
Mech. Eng. 335: Thermodynamics I	3
Mech. Eng. 252: Eng. Mat. and Mfg. Proc	3
or Elec. Eng. 314: Circuit Analysis and Electronics	
Program subjects (30 hours)	
Ind. Eng. 300: Intro. to Operating Systems	3
Ind. Eng. 333: Human Performance	3
Ind. Eng. 310: Operations Res.	3
Ind. Eng. 315: Stochastic Industrial Proc.	3
Ind. Eng. Electives (15 hours—12 to be selected from Ind. Eng. 421, 441, 447, 451, 463, and 473)	15
Technical Electives (15 hours)	
(At least 9 hours must be non-Ind. Eng.)	15
Free Electives (6 hours)	6
Total	128 hours

By the end of the ’60s, there continued to be no specific course requirements for the MS or PhD degrees. The MS degree required candidates to take a total of 30 credit hours (selected from 400- or 500-level courses), 18 hours of which had to be industrial engineering courses. Students entering the master's program had to have completed a BS in industrial engineering or the equivalent. Students who did not have a BS in industrial engineering could be required to take additional courses in statistics, linear programming, or computer programming without graduate credit.

The graduate courses in the department (described later) were clustered into the following categories, each of which had at least two faculty members:

1. Human performance (Hancock and Chaffin)
2. Management engineering (Gage, Johnson, Jelinek, and Steffy)
3. Operations engineering and management science (Thrall, Murty, Disney, Baum, Kimbleton, Galliher. and Bonder)
4. Process systems design (R. Wilson and R. Evans)
5. Computers and information processing systems (Teichroew, Herzog, D. Wilson, Merten, and Sibley)

The PhD students were required to take at least two graduate courses from three of these areas and one graduate course from each of the other two areas. They then had to pass both a written and oral examination of their knowledge in the field. This test was referred to as a steering review and was normally taken at the end of the candidate's second term in residence. Students were also required to demonstrate a reading competency (either by testing or course work) in two foreign languages by the time they graduated, one of which could be satisfied by taking nine credit hours in a cognate area approved by the adviser. With successful completion of the steering review, the PhD students were asked to select members for their doctoral committee and to present a written and oral proposal to them, which represented their thesis preliminary examination. If the proposed thesis was accepted, the student would commence the research and, at some point, present the results in both a written dissertation and a public final thesis defense presentation before the doctoral committee.

2.8 Synopsis of the 1960s—IE Contributions and Major Activities

- IE faculty size doubles from nine to 18 full-time members.
- Sponsored research funding increases from $25,000 in 1962 to $700,000 in 1969.
- A popular off-campus IE master's degree program grows with classes offered in the evening at Flint and Saginaw sites.
- The PhD program increases from about 10 students to more than 40 students.
- The health care systems area, under the direction of Clyde Johnson and Richard Jelinek continues to expand with student projects in several area hospitals.
- The Human Performance Laboratory, under the direction of Walton Hancock, expands its motion and time prediction research to include models of human fatigue and operator learning.

- Space in the North Campus G. G. Brown Laboratory Building is acquired for research offices and laboratories.
- The Defense Systems Laboratory, under the direction of Seth Bonder, completes a number of studies modeling military operations at all levels.

[3]

The Maturing of the IOE Department in the 1970s

3.1 A Difficult Time in Michigan

From about 1968 to 1972 there was a great deal of public outrage over the Vietnam War and the military use of technology for mass destruction. The University of Michigan (UM) saw several student-led riots and public protests over the war, often organized by the Students for a Democratic Society, a national organization with offices in Ann Arbor. In addition, economic and social inequalities between white and black citizens became a dominant public issue. In Michigan this was very much the case after the Detroit race riots in the summer of 1967 that resulted in 43 deaths, 1,189 injuries, and more than 7,200 arrests. This launched the Black Action Movement on campus, which organized several sit-ins and public protests (one of which closed the university for 18 days) and forced university president Robben Fleming in April 1970 to negotiate a plan to increase the number of black students and faculty to 10 percent over the next five years. To accomplish this and meet student demands for more social and humanistic oriented courses, the university immediately began shifting funds away from the sciences and engineering to provide new education and research programs in the social sciences, health care, public policy, and the humanities.

The defunding of science and engineering continued for the better part of the '70s, despite a resurrection of public interest in science and technology that can be traced to the oil crisis that began late in 1973 and lasted until about March 1975. In a few months, the cost of gasoline went from about 25 cents to over a dollar a gallon. This drastic increase was due to the OPEC nations restricting their supply of oil as a means of punishing countries that supported Israel. Many gas stations in the United States and Europe ran out of fuel (imagine long lines at gas stations), which caused the driving public to demand action. After a great deal of political and even military maneuvering, the OPEC countries slowly released more oil into the marketplace throughout 1974, but the cost of oil and gas remained very high. This in turn initiated two technological changes in the United States: First, the development of ways to conserve energy became popular (e.g., lighter and more aerodynamic vehicle designs, more fuel efficient engines, and better insulation in buildings and homes), and second, the search for additional sources of energy (e.g., nuclear power, solar energy, and natural gas) and the development of new technologies to locate and extract oil in the United States. In other words, the public was beginning to see science and engineering as the source of the knowledge needed to deal with the energy crisis.

Unfortunately, the state of Michigan's economic recovery from the oil crisis took much longer, and some would say it is still underway. The reason for this is that for 50 years Michigan had relied on one particular product to drive its prosperity: the automobile. Because of the high cost of gasoline in the '70s, the public shifted to smaller and more fuel efficient cars. This allowed the Japanese and German auto companies in particular to gain a large amount of the US market (think VW Beetle cars and small fuel-efficient engines). Another aspect of the foreign car producers selling in the US market was that by the '80s their cars became recognized as more reliable and of higher quality than those that were being produced by domestic companies. This latter situation was partially the result of the

US auto companies trying to develop many new and different fuel-efficient vehicle designs while having to negotiate new contracts between organized labor, car parts suppliers, and other groups that were needed to improve the quality of their cars. After years of distrust between these groups, negotiating new contracts was difficult. To make matters worse, Iran's supply of oil was cut off when its pro-Western government was overthrown in 1979, causing another, though smaller, oil crisis. By the early '80s the situation in Michigan was so bad that economists recognized that the Big Three domestic auto companies were in a deep recession, and many layoffs and cost-cutting measures were required to preserve their capability to continue producing automobiles in a much smaller segment of the market. Needless to say, the state taxes needed to support the university were eroding with the domestic car market throughout the '70s, precipitating the need to sharply raise tuition to offset declining state support.

The combination of the antitechnology movement in the late '60s and the jobs lost due to the oil crisis in the early '70s had an immediate effect on decreasing the number of engineering students. This is reflected in the decreased number of IE undergraduate degrees being granted, as seen in the graph. On the other hand, recently graduated engineers with BS degrees who were still working or who were recently laid off enrolled in various master's degree programs offered by the College. Since the IE Department was offering master's degree courses in Ann Arbor and Flint, and these programs did not require a master's thesis, enrollment in these programs remained strong throughout the '70s.This was particularly true at UM-Flint, where the courses were offered in the evening. Unfortunately, such was not the case with the PhD program, where the turnover and loss of some important faculty members throughout the '70s, as discussed in the next section, meant a lack of a critical mass of expertise in some of the IE subdisciplines. The result was a steady decline in PhD students completing their degrees until the late '70s, when additional faculty and sponsored research funds once again became available to support the PhD program.

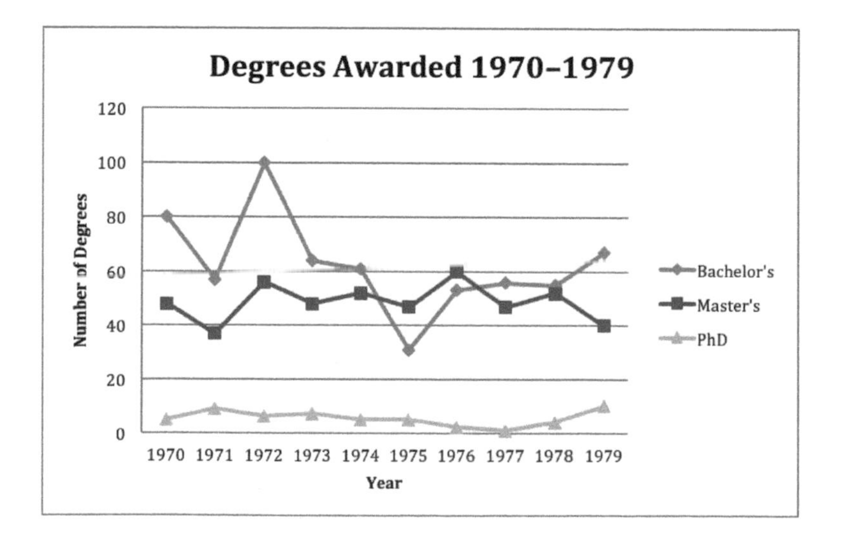

3.2 Building the IE Department in Tough Times

As depicted in the foregoing, the beginning of the 1970s was not good for the IE Department. Faculty turnover was becoming a major issue; turnover occurred partly because salaries were far lower than companies were paying and partly because of what some would say was an unreasonable demand by the administration for new faculty members to quickly find grant and contract funds to support the PhD research program. Also, military-related research, which had supported a number of PhD students during the '60s in the College of Engineering, was under pressure to move off campus, which eventually caused the loss of the IE Department's Defense Systems Research Laboratory in 1972. Because of the declining tax base and the efforts of the university's executive officers to fund socially acceptable areas, including new dental and medical schools (along with a new hospital), the college general fund decreased by about two percent each year throughout the '70s. This resulted in the total number of full-time-equivalent faculty being reduced by almost 25 percent, from 302 in 1970 to 232 in 1979. For a relatively young department that needed to hire and retain more faculty members in the areas believed to have the greatest academic potential, namely computer and information systems, human performance, and operations research, the setting within the College provided major challenges.

Fortunately, at the request of Dean Van Wylen, the leadership of the department had produced a five-year plan in 1969, which provided sound arguments for hiring new faculty members over the next five years in each of the specialty areas of the department. As part of the development of this five-year plan a survey of IE department chairs in the United States was performed. This resulted in data suggesting that the department was now ranked in the top 10 percent of the 48 existing IE departments. But this same survey also raised a question about the low number of publications and lack of visibility of some of the faculty members. Based on these results, the faculty plan made the case that becoming further recognized as a leading department would only be possible if additional research-oriented faculty members were hired in strategic areas, especially since it was expected that some of the senior faculty members who were hired in the '50s would soon be retiring. The resulting five-year plan recommended an increase in the number of full-time faculty members from the existing 18 members in 1969 to 26 members by 1974. This was the plan, but as discussed before, with the decreasing College budgets in the early '70s, the number of full-time professorial IE faculty members had actually diminished further, from 17 in 1972 to 14 in 1975. This was the result of several early retirements, better job offers elsewhere, or denial of tenure in a couple of cases. In fact it was not until 1980 that the department was back to 18 faculty members. It is perhaps worth noting that the reduction in professorial staff meant that many of the 500- and 600-level advanced courses upon which the PhD students relied when preparing for their steering review (later known as the preliminary examination) were not offered. This, combined with reduced PhD fellowship support, had a negative effect on the PhD program in the early to mid-1970s, and many PhD students left the program.

One of the other recommendations of the 1969 IE five-year plan was to expand the space allocated to the department for research, particularly the space in the G. G. Brown Laboratory Building on North Campus. The request was to double the 2,000 square foot space that was now used by IE faculty members for their research programs. Unfortunately, this did not occur until much later, despite the faculty involved in these activities providing more than $200,000 annually (about one-third of the department's funding) in research and training grants during the 70s.

3.3 Hiring Faculty Members in Areas of Need

With the 1969 five-year plan as a guide, the faculty embarked on hiring a number of new faculty members. Among those who had a long lasting effect on the department were the following.

Alan Merten – PhD, Computer Science, University of Wisconsin (1970)

Alan G. Merten was hired in 1970 as an assistant professor with a PhD in computer science from the University of Wisconsin. Merten's expertise was in the use of computer algorithms to facilitate problem solving within complex management systems. He worked with Dan Teichroew in the ISDOS program while in the IE Department. In 1974 he joined the faculty in the School of Business, eventually becoming a professor and associate dean. He left the university in 1986 to become dean of the College of Business Administration at the University of Florida (1986–1989), then dean of the Johnson Graduate School of Business at Cornell University (1989–1996), and, finally, president of George Mason University (1996–2012). (See Marten's web page: http://agmerten.wordpress.com.)

James Miller – PhD, ISE, Ohio State University (1971)

James Miller was hired as an assistant professor in 1971 after receiving his PhD in industrial engineering from the Ohio State University with a specialty in transportation safety research. He established courses in safety systems and safety management, which became an integral part of the newly acquired Safety Engineering Training Grant from the National Institute for Occupational Safety and Health. In 1977, Miller, then an associate professor, took a two-year leave to work for the Occupational Safety and Health Administration in Washington, DC, where he helped establish several national safety regulations. Upon returning to the university, Miller continued to teach courses in safety engineering, while establishing a successful safety consulting company, along with several of his previous PhD students. (See the Seak Expert Witness Testimony website for more on Miller's credentials: www.seakexperts.com/members/3262-james-m-miller.)

Gary Herrin – PhD, IE, Ohio State University (1973)

Gary Herrin was hired in 1973 as an assistant professor after receiving his PhD in industrial engineering from the Ohio State University with a focus on applied statistics and safety systems. Herrin reestablished several statistics and quality-control courses in the department, and became a major leader in statistical models of occupational injury processes. The latter work was commended by the recently established National Institute of Occupational Safety and Health (NIOSH). He served as the founding director of the Center for Ergonomics in 1980, and in 1981 he led a national consensus committee to write the NIOSH *Work Practices Guide for Manual Lifting*, which established the scientific basis for determining the risk of lower-back injury when lifting objects of different size and weight. In the '90s he established a set of video-based lectures on the use of NineSigma methods to improve manufacturing quality control. (For more on Herrin's career, see "In Memory of Professor Gary D. Herrin, PhD, 1946–2011," by Don B. Chaffin: https://ergoweb.com/in-memory-of-professor-gary-d-herrin-phd-1946-2011/, and chapter 6 for more about the Center for Ergonomics.)

Gary Langolf was hired in 1974 after working as an assistant professor in the Industrial Engineering Department at Wayne State University. Langolf had become well known for his studies of perceptual-motor control of movements. His work was critical in minimizing human movement errors when assembling products of high quality. As a member of the Center for Ergonomics, and in cooperation with James Alpers, MD, PhD, a professor in the Department of Neurology, he worked with NIOSH to establish psychomotor tests of workers who were exposed to heavy metal fumes that were suspected of being neurotoxins. This work became the basis for the field of occupational behavioral toxicology. (See chapter 7 for more about the Center for Ergonomics.)

Gary Langolf – PhD, IE, University of Michigan (ca. 1972)

Tony C. Woo – PhD, Electrical Engineering, University of Illinois (1975)

Tony Woo was hired as assistant professor in 1977 to carry on Bert Herzog's pioneering work in computer graphics, a course that was jointly listed between IOE, CICE (later EECS), and much later MEAM. He also taught the rudiments of data processing (e.g., inventory, payment, and invoicing) to all IOE undergraduates. From 1980 to 1987 he served as IOE Department's undergraduate adviser. He served as a National Science Foundation program director and was responsible for three programs from 1987 to 1990. He returned to IOE and was promoted to professor in 1990. His research focused on computational geometry (with applications to computer-aided design [CAD]) and complexity analysis (which is akin to human performance but for algorithms). His work was often pioneering, as there were no CAD programs when he first started, and his publications were highly sought.

In 1995, he left Michigan for the University of Washington, where he held the John M. Fluke chair for manufacturing and was later appointed director of industrial engineering . In 2004, he became dean of graduate studies at Nanyang Technological University in Singapore, where he was later director of admissions for 8,000 graduate students and vice president for research. While at Nanyang he initiated a research scholarship program for the top undergraduates, which allowed participants to do research with any professor in any discipline. He also established an Advanced Institute for Studies, through which Nobel laureates were invited to give lectures and be in residence. He retired in 2007.

Thomas Armstrong – PhD, IE/Physiology/ Occupational Health, University of Michigan (1978)

Thomas Armstrong joined the department in 1991 as a full professor after serving 14 years as a faculty member in the Department of Environmental and Industrial Health in the UM School of Public Health with a joint appointment in IOE. Armstrong holds a multidisciplinary PhD degree in operations engineering, industrial health science, and physiology from UM. He is best known for his work in biomechanics and injury risk associated with handwork. He served director of the Center for Ergonomics from 1998 to 2015. (See more about Armstrong at his UM page: www-personal.umich.edu/~tja/, and chapter 7 for more about the Center for Ergonomics.)

Jack Lohmann – PhD, IE&ME, Stanford (1979)

Jack Lohmann earned his PhD in industrial engineering and engineering management at Stanford University in 1979 and joined the IOE faculty shortly thereafter. He also held visiting appointments at the University of Southern California, l'École Centrale Paris, and the National Science Foundation. In 1987, he was appointed associate dean for graduate and undergraduate studies in the College. He joined the faculty at the Georgia Institute of Technology in 1991. His early career research interests included economic decision analysis and capital budgeting under risk, expanding later to engineering education. More recently, Lohmann was recognized as a pioneer in establishing the field of engineering education research and was credited with elevating the *Journal of Engineering Education* to a premier educational research journal during his tenure as editor. Among the external sponsors of his work were AT&T, Continental AG, Dessault Systemes, ExxonMobil, GM, Hewlett-Packard, IBM, Microsoft Research, Motorola, the National Science Foundation, Procter & Gamble, the Sloan Foundation, and the United Engineering Foundation. Lohmann retired in 2012 as vice provost and professor of industrial and systems engineering at the Georgia Institute of Technology. At that time, his principal responsibilities included accreditation of Georgia Tech's academic programs and serving as the president's liaison to the Commission on Colleges of the Southern Association of Colleges and Schools and the National Collegiate Athletic Association. (See Lohmann's page at ZoomInfo: http://www.zoominfo.com/p/Jack-Lohmann/18820313.)

Other faculty members who contributed for short periods (particularly in the OR-related areas) in the 1970s were Kenneth Baker (1970–1974), Michael Warner (1970–1978), Craig Kirkwood (1974–1980), Frank Noonan (1975–1978), John Bartholdi (1977–1980), Louis Boyston (1977–1983), and Loren Platzman (1978–1980). This period also brought the early retirements of professors Gage, Johnson and Steffy, and professors Jelinek and Bonder left the department to set up and operate successful consulting companies. This loss in senior leadership was compounded when Ed Sibley left to become a department chair at the University of Maryland in 1973, Bertram Herzog became director of the Computing Center at the University of Colorado in 1975, and Ralph Disney accepted a chaired professorship at the Virginia Polytechnic University in 1977.

3.4 Changes in the IE Curriculum

With the hiring of faculty who had expertise in the more theoretical aspects of operating systems during the late '70s, it was possible to expand the number of IE required courses needed to receive a BS degree. The result being that by 1980 the IE department was teaching its own statistics courses, and not relying on the Department of Mathematics. Also, instead of one course in Operations Research methods, two courses were required: One in Stochastic Industrial Processes and the other in Optimal Methods. Lastly, the Human Performance area was able to offer a required human performance laboratory course.

IOE Curriculum in 1980

Course	Credit Hours
Subjects required by all programs (55 hours)	
Mathematics 115, 116, 117, 215, and 216	16
Humanities 1010 and 102: Great Books I and II	6
Engineering 102: Computing	2
Chemistry 123 or 124	3
Chemistry 125	2
Physics 140 with Lab. 141; 240 with Lab. 241	8
Seminars In Literature and Rhetoric	6
Humanities and Social Sciences	12
Related technical subjects (13 hours)	
Appl. Mech. 211: Intro. to Solid Mech.	4
Mech. Eng. 235: Thermodynamics I	3
Mech. Eng. 252: Elem. of Mfg. Sys	3
Elec.-Comp. Eng. 215: Network Analysis	3
or Elec.-Comp. Eng. 314: Circuit Analysis and Electronics	
Program subjects (36 hours)	
I.&O.E. 300: Mgt. of Technical Change	3
I.&O.E. 310: Intro. to Optim. Methods	3
I.&O.E. 315: Stochastic Industrial Proc.	3
I.&O.E. 333 Human Performance	3
I.&O.E. 334: Human Performance Lab	1
I.&O.E. 365: Engineering Statistics	4
I.&O.E. 373: Data Processing	3
I.&O.E. 374: Data Processing Lab	1
I.&O.E. Electives (15 hours)	15
Technical electives (15 hours)	
(At least 9 hours must be non- I.&O.E.)	15
Free electives (9 hours)	9
Total	128 hours

Unfortunately, as indicated earlier, though the master's program remained very popular, the PhD program did not do as well during the '70s. The smaller number of faculty members were able to provide a set of 400- and 500-level courses for the MS degree, but the third-level 600-level courses were not consistently offered. This situation, coupled with reduced support for PhD student research in the OR area, resulted in the PhD enrollment dropping from more than 40 students in 1970 to about 29 students in 1976. The numbers recovered after 1978, and 50 PhD students were enrolled in 1980.

3.5 Organizational Changes—and a New Name

As suggested earlier, it would take the entire decade of the '70s for the public to recognize the importance of technology in their lives and be willing to pay the escalating tuition costs and for the university to urge companies, foundations, and state and federal agencies to provide the research funds necessary to further develop the College. At the beginning of the 1970s, the IE faculty made several important policy decisions to sustain and improve its reputation and capabilities. First was the recognition that the breadth of teaching and research in the department was not well represented by the traditional name "industrial engineering." As noted, some faculty members were already highly involved in problem areas outside of manufacturing, such as hospital systems, health care planning, and transportation systems. At a faculty meeting on March 3, 1971, Stephen Pollock proposed the following alternative courses of action regarding the name of the department:

1. Do nothing.
2. Change the name to industrial and systems engineering.
3. Change the name to industrial and operations engineering.
4. Change the name to operational engineering.
5. Change the name to something else.

The faculty at that meeting voted in favor of a name change, and authorized Pollock to perform a mail ballot to determine what name was preferred. This was done, and at the December 1, 1971, faculty meeting the following motion was made: "Professor Pollock moved that the name of the department be changed to Industrial and Operations Engineering." This motion was seconded and passed. In February 1972 the name change became official with the following action of the Regents:

> The Regents approved the request of the Dean of the College of Engineering to change the name of the Department of Industrial Engineering to the Department if Industrial and Operations Engineering, effective immediately, to better describe the scope of the departmental activities. – (From the February 1972 meeting, *University of Michigan, Proceedings of the Board of Regents [1969–1972]*, page 1347).

It is perhaps worth noting that the new name has some obvious advantages:

- "Industrial" indicates that the world of industry (in particular, manufacturing) remains a major component of the teaching and research mission of the department.
- "Operations" includes phenomena more than those usually associated with industrial situations and challenges. It includes the use of methods to analyze, improve, and direct decision making in such *non*industrial arenas as commercial aviation, health care, medicine, municipal services, distribution and logistics, homeland security, forestry, communications, as well as—not unsurprisingly —areas at the very roots of OR: national defense and other military operations.
- "Engineering" makes it clear where the intellectual foundations of the curriculum and research agendas lie: in a college that not only creates new knowledge via theoretical and applied research but also has an engineering point of view of implementation and evaluation of this knowledge.
- Finally, although the word "research" is missing from the name, it is clearly implied and necessary for an academic institution. But it is not singled out, which suggests that it is not a higher goal than implementation and application (it should be noted that there are few, if any, engineering school departmental names that contain the word "research").

A second major change occurred in the governance of the department. In 1973, the faculty passed a set of bylaws that established a department executive committee. This department committee was composed of four members elected from the faculty to serve for periods of one or two years. The duties were to represent the faculty by advising the chair on hiring and promotion decisions, workload, and general matters pertaining to the department budget and allocation of funds. In essence, the bylaws stipulated that the chair had to vet any matters of importance to the faculty with the department committee before bringing such matters to a vote by the faculty.

A third important management process was also introduced in the early '70s. All faculty members were required to write a short (two-page) summary of their activities over the past year. These "brag sheets" were then circulated to all of the faculty members with a request for them to anonymously comment on and rate the teaching, research, and service activities of each of their colleagues. The chair then used statistical results from the ratings and anonymous comments to prepare annual merit raises and conduct annual performance reviews with each faculty member. Such an open sharing of information about one's activities also allowed all faculty members to better understand what their colleagues were doing and how they might collaborate in the future. The process has continued to this day and has been used by other departments in the College.

A fourth change introduced in the mid-1970s was the development of an effort allocation model. This analytical model was a means of characterizing the time it took to perform various tasks required of all faculty members. Like the use of the annual two-page faculty activity reports, the development of an effort allocation model provided the faculty with information about how their colleagues were allocating their time to perform various tasks. This was deemed especially useful as it became necessary for each one to spend more time performing research and teaching in specialized topic areas within the department and, thus, may not have had the time to teach general required courses, and provide services that related to the larger needs of the whole department. To develop the effort allocation model the faculty had a workshop to determine how much time it would take to perform such tasks as developing a new course; teaching an existing course with and without teaching assistants or graders; advising PhD students; preparing grant and contract proposals; serving on department, college, and university committees; managing a large center or laboratory; conducting sponsored research; and writing scholarly papers. The end result was a linear equation that allowed faculty members annually to use an Excel spreadsheet to list their expected activities for the next year and, by implementing the model, determine if their planned activities would meet the expectations of the group as a whole. The statistical results for all the faculty members were then circulated (without names) so that everyone in the department could better understand what was required of themselves and their colleagues to continue to meet the varied needs of the department. In essence, this management tool allowed everyone to understand how faculty could best distribute their individual time to meet the department's overall goals. In other words, it would allow some to spend more time teaching and would allow others to devote more time to writing proposals or scholarly papers or to advising PhD students. Though not used consistently over the years since it was adopted, it has been helpful in balancing workloads among faculty members and allowing each person to meet personal goals and the goals of other colleagues.

3.6 Rankings Become Important

With about 50 industrial engineering degree-granting programs in the United States by the mid-1970s, the competition for new faculty, research funds, and excellent students drove the need for various rankings of the programs to be published. As was reported earlier, a 1969 survey as part of the IE departmental review revealed that other department chairs ranked the department in the top 10 percent of the 48 current IE programs. As part of the 1977 IOE departmental review, three different public surveys of IE department chairs were cited. One was a study by William Gill from the University of Buffalo, who ranked the UM IE faculty as second only to the faculty at Stanford. A second survey conducted by the *New Engineer* magazine ranked the UM graduate program as the "best academic program," followed by Purdue, Georgia Tech, and Stanford. A third survey by William Biles at Notre Dame, ranked the UM IOE graduate program as third behind Stanford and Cornell. Clearly, despite the difficulties with funding the IE program in the '70s, their peers perceived that the faculty and their programs were performing exceptionally well.

3.7 Synopsis of 1970s—IE Contributions and Major Events

Despite the difficult times for science and engineering and the IOE Department in the early '70s, which resulted in some major turnover in faculty members, there were the following notable contributions and events:

- The Management Information Systems group, composed of faculty members (Sibley, Merten, Herzog, and Woo) under the direction of Dan Teichroew, organized the Information Systems Design and Optimization Laboratory (ISDOS). To fund their research, a consortium of about 20 government and industrial IT companies was established. The laboratory produced a series of reports that detailed the development and use of their Problem Solving Language PSL and their Problem Statement Analyzer PSA. (See chapter 6 for more about ISDOS.)
- The Human Performance and Safety Engineering Laboratory, codirected by Professors Hancock and Chaffin, acquired the first Occupational Safety Engineering Training Grant in 1971 from the new National Institute for Occupational Safety and Health. This training grant supported about six to eight graduate students annually and provided the funds for a set of short courses for practicing engineers interested in learning more about ergonomics and safety engineering.
- In 1979, the regents renamed the Human Performance and Safety Engineering Laboratory as the Center for Ergonomics. Two years later the center involved 15 faculty and staff members from several departments and was supporting 27 graduate students with more than $1 million in sponsored research. (See chapter 7 for more on the Center for Ergonomics.)
- The three faculty members (Johnson, Jelinek and Hancock) in the hospital systems research group continued to expand their student projects in various hospitals. One of the alumni from this group, Karl Bartsch, established Chi Systems, which became a leader in hospital systems consulting services with more than 200 employees (including many IOE graduates). Hancock joined James Martin (IOE-PhD 1974) in the Department of Hospital Administration to offer a master's degree in hospital systems engineering. (See more about the hospital systems area in chapter 6.)
- The Defense Systems Laboratory in IE, established by Seth Bonder in 1962 provided graduate students with the unique knowledge necessary to use a variety of operations research methods in planning military and security operations. In 1972, Bonder founded Vector Research Incorporated and, along with a number of his former graduate students, built it into the leading defense systems analysis company in the country, with more than 400 employees.
- In 1972, the name of the department was officially changed to industrial and operations engineering to better recognize the breadth of teaching and research being conducted by various faculty members.
- The department is ranked among the top three in the United States.

[4]

The Department on the Move—the 1980s

4.1 Things Are Getting Better

Though another short-lived recession hit the auto industry in 1979, by the early 1980s domestic car sales were rebounding and the College of Engineering was beginning to plan major initiatives under the new dean, James J. Duderstadt. One major change that the Duderstadt administration implemented was to reinvigorate construction of new buildings on North Campus, which would eventually house the entire College of Engineering. He and his senior associate dean, Charles Vest, also developed policies and procedures that facilitated the development of major laboratories and centers in the College and developed financial incentives for faculty members to increase their support of PhD students. Finally, the administration was able to provide larger merit pay increases along with public awards for those who had demonstrated outstanding scholarship, teaching, and national service. All of these changes within the College leadership had profound effects on the Industrial and Operations Engineering (IOE) Department in the 1980s.

More specifically, by 1982 the department had 18 full-time faculty members and was offering three specialized master's degree options in occupational health and safety engineering, manufacturing systems engineering, and public systems analysis, in addition to a general industrial engineering (IE) master's degree. These 30 credit hour optional master's degrees required students to take about one-third of their courses from other departments, such as Industrial and Environmental Health, Mechanical Engineering, Hospital Administration, or the Institute of Public Policy, to name a few. At the same time, as these new specialized master's degree programs were being developed on the Ann Arbor campus, the popular evening degree programs being taught in Flint and Saginaw were being phased out, mainly because the younger faculty members wanted to concentrate on building a stronger research program for PhD students on the Ann Arbor campus, which was also consistent with the College administration's interests. This resulted in a decrease in the number of master's degrees being granted starting in 1982, most of which was due to the closing of the evening program at Flint, as discussed in the next section.

4.2 Enrollment Trends

Whereas in the 1970s the average number of IOE degrees granted annually was about 60 BS, 55 MS, and 6 PhD degrees, the average number in the 1980s was much higher for undergraduates (about 100 BS degrees annually) but lower for master's degrees (about 40 MS degrees annually) and about the same for the doctoral degrees (about 6 PhD degrees annually). The graph shows the annual trends in graduation rates. It is important to note that by 1980 enrollment in the required 300-level courses had reached physical capacity (about 80 students) due to the size of the older lecture rooms available in West Engineering. And with a limited number of faculty members, it was decided to not teach multiple sections of these popular undergraduate courses to enable the faculty to continue to rebuild the graduate and research programs. Thus, there was a restriction on the undergraduate enrollments until larger lecture rooms became available after the move to North Campus was completed in 1983.

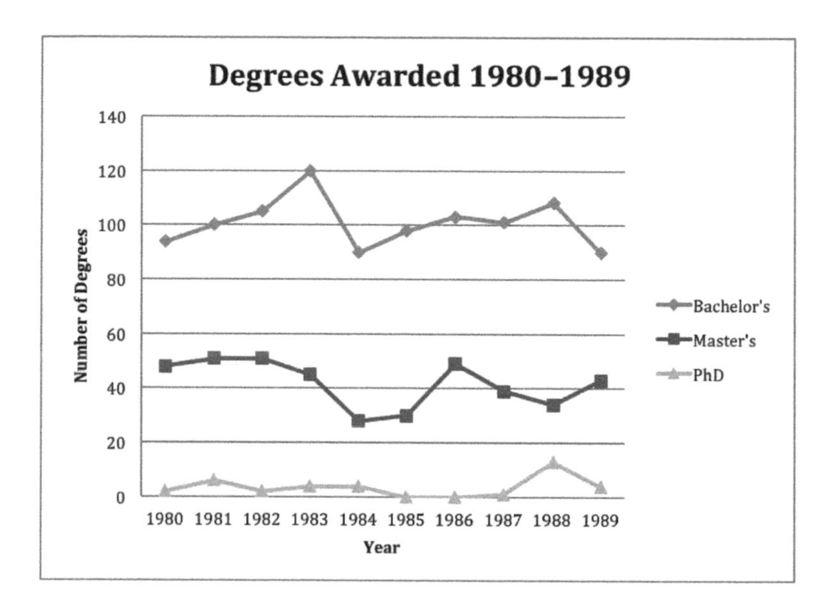

4.3 Faculty Changes

Several very important faculty members had left the IOE Department in the '70s and early '80s. Notable among these in the management information systems area were Bert Herzog, who moved to the University of Colorado in 1975; Anthony Woo, who moved to University of Washington in 1977; and Alan Merten, who moved to the University of Michigan (UM) School of Business in 1976. In the operations research area, Ralph Disney left in 1977 to join the faculty at Virginia Tech University. The loss of these important leaders, along with several other assistant professors who were not promoted or took jobs with higher salaries, meant that the department had to aggressively hire additional people. Fortunately, the following people joined the IOE faculty early in the '80s.

James C. Bean – PhD, Operations Research, Stanford (1980)

James C. Bean was hired in 1980 after completing his PhD in operations research at Stanford University. Bean, an outstanding instructor, taught several courses in the operations research area and became well known for his research on genetic algorithms for highly constrained problems and for system sustainability modeling. He joined with faculty members in the School of Business to form the Tauber Manufacturing Institute, which he codirected in the '90s. This institute supported teams of engineering and business students in tackling a variety of important manufacturing problems in the United States. Prior to that, he joined Bob Smith and Jack Lohmann in the IOE Department to create the Dynamic Systems Optimization Laboratory, which developed analytical models to optimize the performance of operations that have time-dependent conditions. He served as the president of the Institute for Operations Research and Management Sciences, where he was a charter fellow, and later received the George Kimball Medal. At UM he served as the associate dean for graduate education and international programs and then associate dean for academic affairs of the College of Engineering until assuming the position of dean of the Lundquist College of Business at the University of Oregon in 2004. He later served as provost at Oregon (2008–2013) and at Northeastern University (2015–present) (See Bean's page at the UM Center for Sustainable Systems website: http://css.snre.umich.edu/person/james-c-bean .)

Robert L. Smith was hired in 1980 after receiving his PhD degree from the University of California at Berkeley. His research and teaching have focused on dynamic programming and modeling stochastic processes related to communications, traffic routing, and vehicle manufacturing. He served as the director of the Dynamic Systems Optimization Laboratory and has advised 28 PhD students. Smith has held advisory positions at the National Science Foundation and at several other universities, and had been an associate editor of the *Operations Research* and *Management Science* journals. (See his personal page at the UM website: http://www-personal.umich.edu/~rlsmith/.)

Robert L. Smith – PhD, Operations Research, University of California–Berkeley (1980)

John R. Birge – PhD, Operations Research, Stanford (1980)

John R. Birge was hired in 1980 after completing his PhD in operations research at Stanford University. He soon established himself as an expert and scholar in the development and use of stochastic programming and large-scale optimization to solve a variety of important problems in finance, electric power distribution, health care, and vehicle manufacturing with funding from the National Science Foundation, Ford, GM, Volkswagen, the Office of Naval Research, and the National Institute of Justice. He has received awards from ORSA/TIMS and Institute of Industrial Engineers; was president of INFORMS; and was elected to the National Academy of Engineering. In the IOE Department he codirected the Dynamic Systems Optimization Laboratory with Jim Bean and Bob Smith and served as department Chair from 1992 to 1999, at which time he became dean of the McCormick School of Engineering and Applied Science at Northwestern University. (See Birge's page at the University of Chicago Booth School of Business website: http://www.chicagobooth.edu/faculty/directory/b/john-r-birge.)

Jeffrey K. Liker was hired in 1982 to replace Clyde Johnson, who had retired in 1974. Liker received a BS degree in industrial engineering and a PhD in sociology from the University of Massachusetts. One of his initial acts was to work with the faculty in the Center for Ergonomics to study and recommend the type of training that was needed in the auto industry to set up ergonomics programs in all manufacturing facilities. His approach was to have labor and management groups lead these developments, and this has been the foundation for the practice of ergonomics throughout most of the United States. He later conducted a series of studies comparing management styles in the United States and Japan, which resulted in a widely read and critically acclaimed book: *The Toyota Way*. Because of this book and other publications, Liker and his group have received the Shingo Award for Research and Professional Publications three times, which is provided for researchers that advance "new knowledge and understanding of lean and operational excellence." Most recently, he developed and advocated a variety of methods for achieving higher product quality while containing or reducing manufacturing costs. These methods are often referred to as a "Lean Production System." (For Liker's UM web page, see: http://liker.engin.umich.edu/.)

Jeffrey Liker – PhD, Sociology, University of Massachusetts (1980)

Candace Yano – PhD, IE and management sciences, Stanford (1981)

Candace A. Yano joined the department in 1983 after receiving her PhD in industrial engineering from Stanford University. Her primary areas of expertise are production systems, inventory and logistics management, and product attributes that affects marketing. She left the UM in 1993, and from 1995 to 2001 she chaired the Department of Industrial and Operations Research at the University of California at Berkeley. She is a fellow of the Institute of Industrial Engineering and INFORMS. (See her page at the UC-Berkeley website: http://www.ieor.berkeley.edu/People/Faculty/yano.htm.)

W. Monroe Keyserling – PhD, IOE, University of Michigan (1980)

W. Monroe Keyserling was hired in 1984 to expand research and teaching in the area of occupational safety engineering. Keyserling received his PhD in IOE at UM in 1979 and joined the faculty at Harvard University's School of Public Health before returning to UM. His expertise is in ergonomic exposure assessment, epidemiology of work-related injury, prevention of postural fatigue, and safety engineering. He has served as the director of the National Institute of Occupational Safety and Health (NIOSH)–sponsored Center for Occupational Health and Safety Engineering at UM and as graduate program director for the IOE Department. (See Keyserling's web page at the College of Engineering: http://ioe.engin.umich.edu/people/fac/wmkeyser.php and see chapter 6 for more information on the Center for Occupational Health and Safety Engineering.)

Mandyam Srinivasan – PhD, IE and management sciences, Northwestern (1984)

Mandyam M. Srinivasan (Srini) was hired in 1985 after receiving his PhD in industrial engineering and management sciences from Northwestern University. His areas of expertise are global supply chain management and business analytics. He left the university in 1992 and currently holds the Pilot Corporation Chair of Excellence in Business at the University of Tennessee. He received the Franz Edelman Award for Achievement in Operations Research from the Institute for Operations Research and Management Sciences in 2006, the Chancellor's Award for Research and Creative Achievement from the University of Tennessee in 1996, and many awards from the University of Tennessee's Haslam College of Business for outstanding research and teaching and for leadership in executive education. His research and teaching efforts have been supported by grants and contracts from such entities as the US Air Force, the National Science Foundation, Northern Telecom, General Motors, Allied Signal-Honeywell, and IBM. His work has appeared in such journals as *Operations Research*, *Management Science*, *IIE Transactions*, and *Queueing Systems*. Srini worked for many years in two leading automobile manufacturing organizations and successfully installed and managed materials planning and control systems for both. He has consulted with a variety of companies and has published five professional books, the most recent three, published with McGraw-Hill, being *Building Lean Supply Chains with the Theory of Constraints* (2011), *Global Supply Chains: Evaluating Regions on an EPIC Framework* (2013), and *Lean Maintenance, Repair and Overhaul: Changing the Way You Do Business* (2014). (See his web page at the University of Tennessee: http://bas.utk.edu/our-department/faculty/msrinivasan.asp.)

Yavuz A. Bozer was hired in 1986 after receiving his PhD in industrial and systems engineering from the Georgia Institute of Technology. Bozer's teaching and research interests focus on the development of quantitative design and performance improvement/evaluation models and the application of Lean techniques to material flow and storage systems in manufacturing and logistics facilities. In 1988, he was named a Presidential Young Investigator by the National Science Foundation, and he received the Technical Innovation Award in Industrial Engineering from the Institute of Industrial Engineers in 1999. He is also a coauthor of *Facilities Planning* (Wiley), a well-regarded textbook that has been translated into Chinese, Japanese, Korean, and Spanish and revised numerous times. Bozer has served for 11 years as the engineering codirector of the Tauber Institute for Global Operations and the codirector of the UM Lean Manufacturing and Lean Logistics certificate programs. (See Bozer's UM web page: http://ioe.engin.umich.edu/people/fac/yabozer.php.)

Yavuz A. Bozer – PhD, Industrial and Systemes Engineering, Georgia Tech (1986)

Romesh Saigal – PhD, OR, University of California–Berkeley (1968)

Romesh Saigal joined the department in 1986 after receiving his PhD from the University of California at Berkeley in 1968. Prior to coming to the UM, Saigal served on the faculty at the UC Berkeley School of Business and at the Northwestern IEMS Department and worked at the Bell Telephone Laboratory. Saigal teaches courses in continuous optimization, linear programming, and financial engineering. His current research involves understanding risk in operational settings within the application areas of transportation, health care, and finance. He has been an associate editor of *Management Science* and is a member of Society of Industrial and Applied Mathematics, American Mathematical Society, and American Association or the Advancement of Science. (See his UM web page: http://ioe.engin.umich.edu/people/fac/rsaigal.php.)

Other faculty members who were with the department for short periods during the 1980s were Devinder Kochhar (1980–1986), William Kelton (1983–1986), Jay Elkerton (1986–1990), and Terrence Lam (1988–1994). Also, David Kieras from the Electrical Engineering and Computer Science Department held a joint appointment and taught courses in human factors and ergonomics from 1984 to 1993.

4.4. The Undergraduate Curriculum and Graduate Program Areas

The undergraduate curriculum did not change much during the '80s. The laboratory course in data processing was combined with the lecture course to provide a four-credit-hour data processing course. Also, a specific number of credit hours were set for IOE and non-IOE technical electives. These changes are shown in the IOE Curriculum in 1990 table.

With the full-time faculty averaging 20 members from 1984 to 1990, the graduate-level 500 and 600 courses were now being offered on a regular basis, particularly in operations research (OR) and ergonomics, with six full-time faculty in each of these areas. The four other areas—information systems, applied statistics, engineering economy and management engineering, and production and manufacturing engineering—had a minimum of two faculty members each. However, there was a growing concern among the faculty that these latter four areas did not have a sufficient number of faculty members for UM to be recognized as a national leader in those disciplines. This led to a recommendation in the 1986 departmental review that the department should be reorganized into three areas, not six, as depicted in a diagram from the report.

IOE Curriculum in 1990

Course	Credit Hours
Subjects required by all programs (56 hours)	
Mathematics 115, 116, 215, and 216	16
English 125: Intro. Composition	4
Engineering 103: Computing	3
Chemistry 130	3
Chemistry 125	2
Physics 140 with Lab. 141; 240 with Lab. 241	8
Senior Technical Communication 498	3
Humanities and Social Sciences	17
Related technical subjects (13 hours)	
Mech. Eng. 211: Intro. to Solid Mech.	4
Mech. Eng. 235: Thermodynamics I	3
Mech. Eng. 282: Elem. Mfg. Sys	3
EECS 314: Circuit Analysis and Electronics	3
Program subjects (36 hours)	
I.&O.E. 300: Mgt. of Technical Change	3
I.&O.E. 310: Intro. Optim. Methods	3
I.&O.E. 315: Stochastic Industrial Proc.	3
I.&O.E. 333 Human Performance	3
I.&O.E. 334: Human Performance Lab	1
I.&O.E. 365: Engineering Statistics	4
I.&O.E. 373: Data Processing	4
I.&O.E. 374: Senior Design Course	3
I.&O.E. Electives (12 hours)	12
Technical electives (15 hours)	
(6 hours must be I.&O.E.)	6
(9 hours must be non-I.&O.E.)	9
Free electives (8 hours)	8
Total	128 hours

1986 DEPARTMENTAL REVIEW
CONCEPTUAL OVERVIEW OF I.O.E.

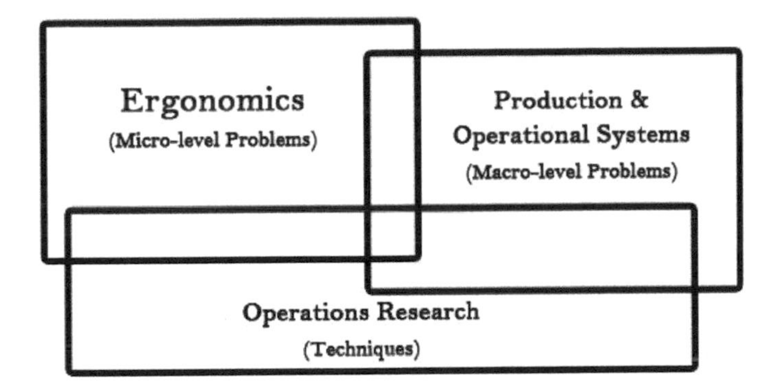

Ergonomics
- Human performance
- Work measurement
- Biomechanics
- Safety engineering
- Legal issues
- Human-Computer interaction

Operations Research
- Mathematical programming
- Stochastic processes
- Reliability
- Decision Analysis
- Statistical Modeling
- General Operations Research
- Simulation

Production & Operational Systems
- Inventory, material & production systems
- Engineering economy
- Hospital systems
- Public policy systems
- Organizational design and management
- Statistical quality control
- Manufacturing systems
- Information systems
- Computational geometry
- Computer graphics

Proposed reorganization of the IOE Department (from the 1986 departmental review).

The 1986 reviewers argued that by combining the four areas that did not have a "critical intellectual mass" the faculty members so involved could select (or hire) a senior faculty member that would represent their collective interests in such matters as fund raising, peer reviews, student recruiting and advising, and space planning. Though conceptually this seemed reasonable at the time, and in fact was strongly supported by two of the external reviewers involved in the 1986 departmental review, such an amalgamation did not occur. The department continued to depict itself as having six areas of concentration, with the courses listed in the chart as being provided on an annual or semiannual basis. The faculty also continued to offer PhD steering examinations in these six areas.

1984 List of Industrial and Operations Engineering Courses by Area

(cross-listed departments are shown in parentheses)

Applied Statistics

IOE 365 (Stat 311) – Engineering Statistics

IOE 460 – Decision analysis

IOE 465 – Design and Analysis of Industrial Experiments

IOE 466 (Stat 466) – Statistical Quality Control

IOE 560 (Stat 550/SMS 603) – Bayesian Decision Analysis

IOE 562 – Multi-objective Decision

IOE 565 – Forecasting and Time Series Analysis

Engineering Economy and Management Engineering

IOE 300 – Management of Technical Change

IOE 421 – Work Organizations

IOE 451 – Engineering Economy

IOE 481 – [formerly IOE 495] Special Projects in Hospital Systems

IOE 503 (EECS 509) – Social Decision Making

IOE 522 – Theories of Administration

IOE 551 – Capital Budgeting

IOE 563 – Labor and Legal Issues in Industrial Engineering

IOE 581 – Hospital Systems Engineering

Ergonomics

IOE 333 – Human Performance

IOE 334 – Human Performance Laboratory

IOE 432 – Industrial Engineering Instrumentation Methods

IOE 433 (EIH 656) – Occupational Ergonomics

IOE 439 – Safety Management

IOE 463 – Work Measurement and Prediction

IOE 533 – Human Factors in Engineering Systems I

IOE 534 (BIOE 534) – Occupational Biomechanics

IOE 539 (EIH 635) – Occupational Safety Engineering

IOE 633 – Man-Machine Systems

IOE 635 (BIOE 635) – Laboratory in Biomechanics and Physiology of Work

IOE 639 – Research Topics in Safety Engineering

Information Systems

IOE 373 – Data Processing

IOE 473 – Information Processing Systems

IOE 478 (EECS 487) – Interactive Computer Graphics

IOE 484 (EECS 484) – [formerly IOE 577] Database Management Systems

IOE 564 (ME 564) – Computer Aided Design Methods

IOE 573 – Analysis, Design, and Management of Large-Scale Administrative Information Processing Systems

IOE 575 – Information Processing System Engineering

IOE 578 (EECS 588) – Geometric Modeling

Operations Research

IOE 310 – Introduction to Optimization Methods

IOE 315 – Stochastic Industrial Processes

IOE 416 – Queueing Systems

IOE 472 – Operations Research

IOE 474 – Simulation

IOE 479 (IPPS 479) – Operations Research for Public Policy

IOE 510 (Math 561/SMS 518) – Linear Programming I

IOE 511 (EECS 503/Math 562/AE 577) – Continuous Optimization Methods

IOE 512 – Dynamic Programming

IOE 515 – Stochastic Industrial Processes

IOE 574 – Simulation Analysis

IOE 610 (Math 663) – Linear Programming II

IOE 611 (Math 663) – Nonlinear Programming

IOE 612 – Network Flows

IOE 614 – [formerly IOE 514] Integer Programming

IOE 616 — [formerly IOE 516] Queueing Theory

IOE 640 — [formerly IOE 540] Concepts in Mathematical Modeling of Large-Scale Systems

IOE 645 — [formerly IOE 545] Reliability, Replacement, and Maintenance

IOE 712 – Infinite Horizon Optimization

Production and Manufacturing

IOE 424 – Production and Service Systems

IOE 441 – Production and Inventory Control

IOE 447 – Facility Planning

IOE 471 (ME 483) – Computer Control of Manufacturing Systems

IOE 494 (EECS 467/Me 484) – Robot Applications

IOE 541 – Inventory Analysis and Control

IOE 543 – Theory of Scheduling

IOE 547 – Plant Flow Analysis

IOE 641 – Seminar in Production Systems

Of note, the 1990 departmental review specifically refuted the 1986 recommendation regarding the consolidation of the four smaller subgroups in the department. The 1990 reviewers believed the breadth of activities conducted by these four groups and their collaborations with other units in the University provided intellectual strength within the department. They believed there was considerable vitality and individual scholarship within each of these four areas and, thus, recommended that each be allowed to hire new faculty members. This recommendation provided the basis for some of the hiring that was undertaken in the '90s.

4.5 Space Wars

As mentioned before, by 1980 the faculty was confronted with two big issues related to space. First, every day faculty members and students who were doing laboratory-based research had to drive between classrooms in the West Engineering Building in Central Campus and the North Campus, where by then almost 4,000 square feet of office and laboratory space was available in the G. G. Brown Laboratory Building. Here they were surrounded by faculty members from other departments, and although that promoted some sharing of resources, it also meant the IOE rooms were not contiguous and were sometimes located on different floors.

The second space problem, commented on earlier, was the relatively small size of the lecture rooms in the old West Engineering Building. Not only were the rooms inadequately ventilated at times (no air conditioning) but the small size of the rooms actually required a cap to be placed on the department enrollment, a problem as undergraduate enrollments surged in the late '70s and early '80s.

So in 1979, the departments that had significant activities on both campuses began a series of discussions on how to solve the problem. One suggestion was to modify the rooms in West Engineering and then realign the occupancy of these so that a couple of departments would move some or all of their activities on North Campus back to West Engineering, while others would move from West Engineering to North Campus. With the relatively small amount of space IOE was occupying on North Campus, many thought the IOE group would be the easiest to consolidate in a renovated West Engineering.

After crunching the numbers on the cost of renovation and moving the IOE labs, however, another much more appealing option developed in 1981. Under the leadership of Dean James Duderstadt and Senior Associate Dean Charles Vest, construction of additional engineering buildings on North Campus would commence, and some of the nonengineering activities would move back to Central Campus. For example, one of the buildings on North Campus was occupied by the University's Research Administration Department, which only required offices, so they could be easily relocated to West Engineering, bringing them closer to the University administration. With modifications to this North Campus building it became feasible for all of the IOE faculty and laboratories to complete the move into the building in November 1983. The building now contained faculty and staff offices, a small library, and seminar rooms on the first and second floors. The basement housed the ergonomics laboratories, a shop, and a large, shared office for PhD students. There was one existing problem, however; except for a few small lecture rooms and laboratory teaching areas, the building did not provide space for the large 300-level courses. The answer to this was the new Dow Chemical Engineering Building, which opened in the winter of 1983. This building had several lecture rooms that could accommodate more than 100 students and was an easy walk from the IOE Building.

Now the faculty could allow more undergraduates into the department and could provide a much more accommodating environment for the PhD students and their research.

The IOE Building in 1983. The open field in the foreground is now occupied by the IOE wing that houses the Center for Ergonomics offices, shops, IOE lecture rooms, and a reflecting pool.

The front entrance to the IOE Building in 1983.

4.6. Research Activities Begin to Shift

As mentioned before, the department had two formal research groups in the '70s: The Information Systems Design and Optimization Systems (ISDOS) group, which was formed by Dan Teichroew when he came to UM in 1968, and the Center for Ergonomics, which was approved by the regents in 1979. The ISDOS organization produced a series of reports and software that allowed information systems analysts to specify the requirements and capabilities of a proposed information technology (IT) system in such a way that the software algorithms would provide guidance on how the system should be designed to best meet the system objectives. At its peak in the late '70s ISDOS employed more than 15 graduate students and had annual funding of about $1.1 million from a variety of companies and government organizations. In 1983, Teichroew and the sponsors deemed the software and associated technical documents to be mature enough to support a new software company, ISDOS Inc. At this time it was decided that the research and development work would remain in the IOE Department as the Program for Research on Information Systems Engineering (PRISE), and most of the ISDOS staff would move off campus to form the new company. Unfortunately, by 1985 only two IOE faculty members were performing research related to information systems engineering (Teichroew and Woo), and each had different interests. Though ISDOS Inc. continued to provide useful IT services to a number of companies, the support of the research program within PRISE dwindled, and by the early 1990s, PRISE no longer existed. Part of the reason for this was that other academic organizations at UM and elsewhere were emerging as computer science or computer engineering departments, leading to less demand and support for this area within IOE. Another reason for the diminishing support for PRISE from ISDOS Inc. was that the UM attorneys believed there existed a conflict of interest on the part of Teichroew because he was running a for-profit company and should not mix funding for its services and his UM PRISE research. Fortunately, the discussions around the latter issue over the years has resulted in a much more supportive policy for those faculty members today who wish to be entrepreneurs while continuing their research within the university. (See chapter 6 for more about ISDOS.)

The second formalized research organization within the department was the Center for Ergonomics. Its purpose was to bring together researchers from different departments to study problems that arise when an engineered system does not have appropriate human-hardware interfaces. The center became recognized as the best such organization in the country, and 15 faculty members and more than 30 PhD students from different departments were involved annually throughout the '80s and '90s. During the period from 1980 to 1990, 29 PhD degree recipients and 105 MS degree recipients did their research work in the center.

Collaboration on ergonomic research problems across several departments was enhanced when the earlier 1971 training grant obtained from NIOSH in occupational safety engineering was combined in 1982 to be part of the NIOSH sponsored UM Educational Resource Center in Occupational Health and Safety Engineering. This multimillion dollar grant supported faculty members and graduate students in industrial and operations engineering, industrial hygiene, occupational medicine, and occupational health nursing. The annual funds provided by the NIOSH grant and other research sponsors to support the Center for Ergonomics research and education programs throughout the '80s was about $1.3 million. (More details about the Center for Ergonomics and the NIOSH training grant are presented in chapter 7.)

As mentioned earlier, the development of a strong OR group within industrial engineering departments began to happen shortly after World War II. Within UM's IOE Department this trend accelerated in the '80s with the hiring of several OR oriented faculty members, in particular, Birge, Bean, Saigal, and Smith. These OR faculty members, along with Katta Murty and Steve Pollock, established through their books and research papers that their OR methods could be applied to resolve a large array of important problems, such as how often car emissions should be tested to meet environmental requirements, the use of Markov decision methods for planning new products, single machine maintenance scheduling based on random breakdowns, the effects of capital budgeting for dealing with risk, and a new method for solving the set partitioning problem. From 1985 to 1990 the OR faculty members and their PhD students published 87 refereed papers and four books. They also advised 11 PhD graduates and served on several editorial boards and in national leadership positions. In 1983, they formed the Dynamic Systems Optimization Laboratory, which is described in chapter 6.

One of the smaller areas of concentration in the department was the production and distribution systems group. It was composed of four faculty members, Bozer, Srinivasan, Wilson, and Yano, who supervised six PhD graduates and published more than two dozen refereed papers over that same period.

Another small area of concentration in the department that continued to be important in the instructional program and in its research findings was the engineering economy and management engineering area. It was composed of faculty members Hancock, Liker, and Lohman, who advised seven PhD graduates and published 28 refereed papers from 1985 to 1990.

4.7. National Reputation Continues to Be Near the Top

All of the activities during the '80s enhanced the national reputation of the IOE Department. The 1986 departmental review cited the 1983 Gorman Report, a national survey of department chairs, which ranked the UM IOE Department as number one in graduate education (UC-Berkeley and Stanford were second and third, respectively). This same report placed the undergraduate program second behind Stanford. According to the 1990 departmental review, the 1989 Gorman Report continued to rank the graduate program number one and the undergraduate program as second to Stanford. The much more analytical 1990 *U.S. News & World Report* list of "Best Engineering Departments" placed the IOE Department in third place behind the larger Georgia Institute of Technology and Purdue University IE departments, which were in first and second place, respectively.

4.8. Synopsis the 1980s: IE Contributions and Major Events

- The size of the faculty averaged about 20 members, and this provided the department with the ability to maintain an undergraduate program that annually granted more than 100 BS degrees as well as a graduate program that granted about 40 MS degrees and seven PhD degrees.
- The importance of the information systems area within the department diminished with the loss of several IT-oriented faculty members and as computer science and computer engineering departments were formed at UM and elsewhere.
- The Center for Ergonomics expanded to include faculty from a variety of departments. With annual funding of more than $1.3 million, it was able to support 25 to 30 PhD students from a variety of departments each year.
- The entire IOE Department moved to the University's research administration office building on North Campus in 1983.
- Various national surveys of IE departments placed the IOE Department first or third, depending on the process used in conducting these surveys.

[5]

A Period of Growth and Change—1990–2005

5.1. National and International Events Affect the Department

The 15 years between 1990 and 2005 provided challenges and opportunities for the Industrial and Operations Engineering (IOE) faculty, and many world events had a direct bearing on the department. The US population had increased by almost 10 percent from 1980 to 1990, some of this attributed to a longer life expectancy, and increased again by more than 13 percent between 1990 and 2000. This population expansion and longer longevity also increased the need for health care service.

The first area of opportunity related to health care. By 2000, the US cost of health care per capita, which was already higher than that of any other industrialized country, was consuming more than 12 percent of gross national product. The high cost of health care provided an incentive for federal agencies, insurance companies, and large hospitals to establish cost containment programs, and these often relied on industrial engineering graduates to develop and implement. The long-standing IOE program in hospital systems engineering benefited from this growing concern. (See chapter 6 for more information about this program.)

The second area related to the military. Although the early 1990s brought an end to the Cold War, it also brought conflicts in other parts of the world, including Iraq's invasion of Kuwait in 1991, which led to the Gulf War. (As a side note, much of the logistics planning for Operation Desert Storm was developed by Vector Research Inc., a company established in the '70s by IOE professor Seth Bonder and a number of IOE alumni). In the next decade, the September 11, 2001, terrorist attacks on the World Trade Center and the Pentagon precipitated the war on terror and led to US troops in Afghanistan and Iraq. As a result, military spending greatly increased, including funding for basic and applied research to improve soldier systems, logistics, and military asset scheduling and planning methods, areas in which IOE has applications.

Environmental issues also dominated the news. The period from 1990 to 2005 saw major hurricanes (Andrew in 1992 and Katrina in 2001) and other natural disasters ranging from massive storms to floods, heat waves, and blizzards, all of which caused deaths and billions of dollars in damage. Scientific evidence of climate change had led to growing public awareness of the issue. In turn, this generated more funding for and student interest in environmental programs of all kinds. In engineering this stoked interest in developing environmentally friendly product designs and manufacturing systems and led organizations to discuss how to design sustainable systems with a minimal long-term adverse impact on the environment.

During the 1980s there was also a growing interest in large corporations to ensure that their managers understood how to use technology to improve the performance of their business. In 1991, the College of Engineering and the School of Business launched a cooperative effort, the Michigan Joint Manufacturing Institute, to encourage students and faculty members to develop joint educational programs related to improving manufacturing systems. This later became the Tauber Manufacturing Institute (TMI), which coordinates joint engineering degrees and summer student project teams between the two schools. By 1996, more than 100 students were enrolled in its master's degree programs. (See chapter 6 for more about this important program, which has been lead by several IOE faculty members.)

The '90s also brought a growing realization, especially in large corporations, that markets are global, and to serve these international markets, products and services had to be developed that would meet the needs of people from a variety of cultures. Work by respected economists, such as the 2005 book *The World Is Flat* by MIT's Thomas Friedman, stressed the need for engineers to gain a broad understanding of different cultures and international business practices. In the IOE Department this type of thinking led to the development of the Engineering Global Leadership (EGL) program 1992. The resulting five-year, BS/MS program required courses offered in both the College of Engineering and the School of Business, a summer internship, and study abroad. Only select students with high grade point averages were admitted into the EGL program, and thus it became the first honors program in the department and in the College of Engineering. By 1996, 35 IOE students were enrolled in the EGL program. (More information about the EGL program is presented in chapter 6.)

Another example of how global issues were affecting engineering was realized from the work of IOE professor Jeff Liker, who had been studying the engineering and management practices of Japanese automotive company Toyota. His studies contrasted the management and engineering processes used by Toyota with those of several US companies. He authored several popular books on this topic and in 1993 began offering short courses that presented fundamental principles related to engineering production systems. In his books and lectures he stressed the importance of product quality and the elimination of waste in all facets of the planning and management of a production system. By this time Toyota had become the world's largest automobile manufacturer; thus, Liker's writings and teaching about their production systems were very popular. (See Appendix A.5 for a list of Liker's books on this subject.)

Also in 1993, the chair of the IOE Department, Chelsea White, organized and directed the Intelligent Highway Systems Center, which was funded by the state of Michigan and several automotive companies. This program combined courses from IOE, electrical and computer science, the School of Business, and the School of Urban Planning to provide an interdisciplinary master's degree. The interest in designing safer and more efficient transportation systems in the United States had great public appeal, and by 1996 this MS program had 65 students.

Lastly, by 2000 the stock market was at its all-time high. Some of this success was credited to the sophisticated analytical models that were being used by large stock brokerage firms on Wall Street. These stochastic and statistically based models predicted the probability that certain companies would meet or not meet their financial goals in a given time period, and thus proclaimed with statistical confidence that a company posed a good or bad risk for investment. The excitement over the development and use of such analytical models became the basis for financial engineering degree programs in several industrial engineering and operations research programs in the United States. Within the IOE Department this interest guided the hiring of several new faculty members with expertise in this area and the development of a multidisciplinary master's degree program in financial engineering, first led by IOE professor John Birge. (See chapter 6 for more information about this program.)

5.2. Enrollment Trends and New Space

As a result of all of these national trends, the IOE Department experienced a large increase in student enrollments. This was particularly true at the undergraduate and master's levels, as depicted in the graph showing the annual degrees granted by the department. In fact, a 2001 survey of industrial engineering department heads indicated that the popularity of the undergraduate IOE program at the University of Michigan (UM) resulted in twice the number of students enrolled compared with other industrial engineering departments at the time. This same survey also reported that more than 100 students had enrolled in the IOE master's program in the preceding year, almost twice the number of MS students per faculty member at any of the other schools included in the annual survey.

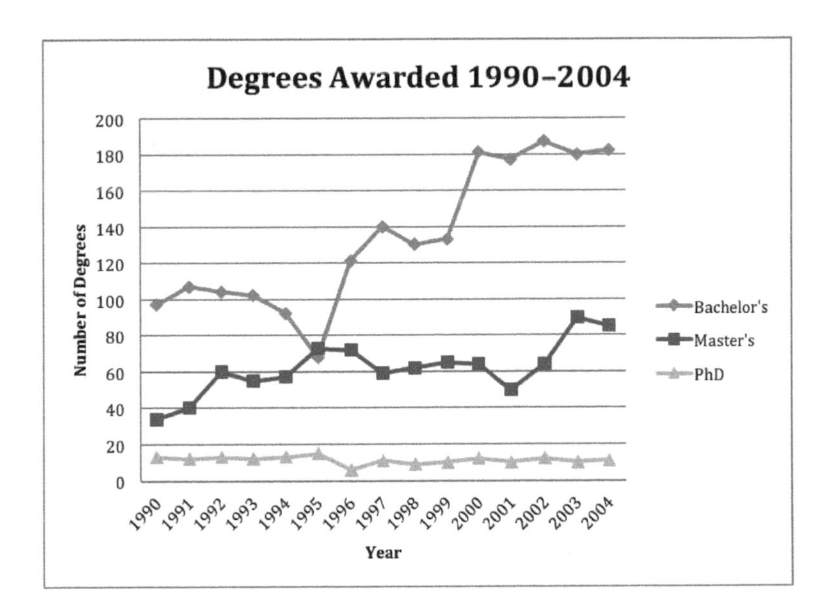

It is interesting to note that the dip in undergraduate degrees awarded from 1993 to 1995 coincided with a period of changes in the IOE physical facilities. The IOE building on North Campus was being transformed and expanded by about 16,000 square feet, which included additional offices, laboratory spaces, and most of all, contiguous lecture and seminar rooms. At the same time the Lurie Engineering Administration Building was being built within a couple hundred feet of the IOE building. The construction around and in the existing IOE building provided some challenges for the students, staff, and faculty for a couple years. When the construction was completed in 1996, the new IOE building was much more welcoming for all students, and this, along with the hiring of some exciting new full-time and adjunct faculty members and the development of various master's degree programs, resulted in the undergraduate and graduate programs becoming very popular in the late '90s and beyond.

The entrance to the new IOE building in 1996. The original building and entrance are behind the trees on the left. The Lurie Engineering Administration Building is partially shown on the right.

5.3 Improving the Computing Environment

By the early '90s it was clear that networked personal computers would be providing the computational capabilities that previously relied on large mainframe computers in a computer center. The early establishment of the Computer Aided Engineering Network (CAEN) in the 1980s, led by Associate Dean Dan Atkins, not only facilitated the use of personal computers by faculty and PhD students for research purposes but also provided personal computer laboratories where students at all levels could use high-level personal computers to solve routine homework problems and perform research on a variety of problems. The CAEN program in the College also provided, at little or no charge, contemporary software applications in a number of engineering domains. This latter aspect of the information technology (IT) environment in the '90s required departments throughout the College to hire computer hardware and software support staff who were able to install and maintain the large array of applications students and faculty demanded. The IOE Department was no different in this regard. Soon, the department had established a small but very capable staff of software experts, led by Christopher Konrad, a UM graduate with degrees in IOE and computer science, assisted by Rod Capps and Mint Minto. The IT support that this staff provided, combined with the specialized hardware and software computing staff support embedded in the Center for Ergonomics, lead by James Foulke and Charles Woolley, assisted by Eyvind Claxton, provided an excellent environment for the development of original computer algorithms to solve a vast array of IE problems. (More on this aspect of the department is discussed in chapters 6 and 7.)

Christopher Konrad – Manager, IT and Facilities

James Foulke – Ergonomics Research Engineer

Charles Woolley – Ergonomics Research Engineer

Rod Capps – Web Applications Developer

5.4 Rankings Remain High

Eyvind Claxton – Engineering Technician

All the new developments in the educational programs during the '90s, the new teaching and laboratory spaces, and the hiring of excellent new faculty and staff resulted in the 1996 Gorman Report ranking the department's graduate program as number one in the country; *US News and World Report* ranked the undergraduate program number one and the graduate program number three. A 1995 National Research Council survey of doctoral programs ranked the doctoral program in IOE as number two in the United States, behind the much larger School of Industrial and Systems Engineering at Georgia Tech University. The 2003 *US News and World Report* ranked the department number two overall. According to two departmental reviews (performed in 1996 and 2003), most agreed that the operations research and ergonomics areas of the department remained the driving force behind these high rankings. It also is worth noting that in 2001 the annual survey of industrial engineering department heads in the United States indicated that the Michigan faculty had the largest number of members who had achieved fellow status in various scientific and professional societies. (See Appendix A.3 for a listing of some IOE faculty awards and honors.)

5.5 Faculty Changes

From 1990 to 2005 the department hired 15 new faculty members who had a major effect on the department, but lost 12 to other universities or to retirement. In addition, 10 faculty members hired during the period were with the department for six or less years. The 2003 departmental review indicated that from 1990 to 1996 there were 23 tenure-track faculty members, and the full-time equivalent (FTE) count remained steady at about 18.5 members on the General Fund budget. Unfortunately, there was a lag in replacing some of the faculty members who left during the next few years, and by 2001 the FTE count was down to about 14.3 members. With the rapid rise in undergraduate enrollments in the department during this same period, however, the department was able to quickly hire more faculty members, and by 2005 the FTE was back to 19 members, and 25 faculty members were in tenure track positions. What follows is a brief description of the 15 faculty members hired during this period who had a significant and lasting effect on the IOE Department.

Chelsea White, PhD

Chelsea (Chip) White joined the IOE Department in 1990. He had received his PhD from the UM in 1974 in computer information and control engineering. He then served on the faculty at the University of Virginia until returning to UM. In 1993, he chaired the IOE Department. He also developed a consortium of transportation companies that combined state and federal transportation funds to establish the Intelligent Highway Systems Center. White served as president of the IEEE Systems, Man, and Cybernetics Society in 1992, and received several IEEE Awards for his "significant contributions in research developments in global transportation and logistic systems". He is an expert in the use of real-time information to improve logistic decisions of all types. He left UM in 2002 to take an endowed professorship at the School of Industrial and Systems Engineering at Georgia Tech University, and in 2005 became chair of the school and director of their Trucking Industry Program. (See White's page on the UM Faculty History Project website: http://um2017.org/faculty-history/faculty/chelsea-c-white-iii.)

Bernard Martin joined the IOE Department in 1990. In 1981, he received his PhD in neuroscience from the Universite' d;Alix-Marseille in France. Martin worked as a research scientist at the National Transportation Research Institute in Lyon, France, until joining the department. His expertise is in neuromuscular control of movements and the modeling and prediction of manual exertions that cause muscle fatigue. His work provides guidance in the design of hand tools, keyboards, and other devices commonly used in workplaces. (See Martin's College of Engineering web page: http://www-personal.umich.edu/~martinbj/.)

Bernard J. Martin, PhD

Yili Liu, PhD

Yili Liu joined the department in 1991 after completing his PhD in engineering and cognitive psychology at the University of Illinois. His research interests include human information processing, cognitive engineering, and cognitive ergonomics, and his research focus is on computational cognitive modeling. He has developed a queuing network model of mental architecture for modeling performance of complex cognitive tasks. This work has provided a quantitative basis for the evaluation and design of a variety of human-machine system interfaces. In 2003, he became an Arthur

F. Thurnau Professor. Liu has received more than a dozen teaching awards and is the coauthor of a best-selling human factors textbook. He has been serving as the IOE undergraduate program adviser for almost two decades. (See Liu's UM web page: http://www-personal.umich.edu/~yililiu.)

Vijay Nair, PhD

Vijay Nair joined the departments of statistics and IOE in 1993 as a full professor. He received a PhD in Statistics from the University of California–Berkeley, in 1978 and then joined Bell Laboratories in their Operations Research Center and later moved to their Mathematical Sciences Center. In 1998, he became chair of the Department of Statistics and, in 2002, received a collegiate-chair professorship. His research interest is quality engineering, especially industrial experiments, reliability engineering, and process control. He is a fellow of the America Association for the Advancement of Science, the American Society for Quality, the American Statistical Association, and the Institute of Mathematical Statistics. (See Nair's UM web page: http://dept.stat.lsa.umich.edu/~vnn/.)

Chien-Fu Jeff Wu PhD

Jeff Wu came to the University of Michigan as the H. C. Carver Professor of Statistics and an IOE professor in 1993. He received his PhD in statistics from the University of California–Berkeley, in 1976 and then held faculty positions at the University of Wisconsin and the University of Waterloo before joining UM. He was elected to memberships in the National Academy of Engineering (2004) and Academia Sinica (2000). He has fellow status in the Institute of Mathematical Statistics (1984), the American Statistical Association (1985), the American Society for Quality (2002), and the Institute for Operations Research and the Management Sciences (2009). Wu is highly regarded as an expert in the development and use of advanced statistical methods to solve product quality and reliability problems. His work has garnered him major awards and honors from various scientific and professional societies. He served as chair of the Department of Statistics from 1995 to 1998. In 2004 Wu, became the Coca Cola Chaired Professor in the School of Industrial and Systems Engineering at Georgia Institute of Technology. (See Wu's Wikipedia entry: https://en.wikipedia.org/wiki/C.F._Jeff_Wu.).

Jan Shi completed his PhD in mechanical engineering at UM in 1992 and then joined the Mechanical Engineering Department. In 1995, he became a tenure-track faculty member in the IOE Department and was later appointed the G. Lawton and Louise G. Johnson chaired professorship in IOE. Shi's research interests focus on system informatics and control for the design and operational improvements of manufacturing and service systems. He has published one book and more than 160 frequently cited papers. He has worked closely with many different industrial companies, has served as principle investigator and co–principle investigator for $19 million of research grants and has advised 28 PhD students. Shi has received many honors for his research from the Institute of Industrial Engineers and other professional and research societies. The technologies developed in his research group have been implemented in various production systems and have had significant economic impacts. (See Shi's web page at Georgia Tech: http://www2.isye.gatech.edu/~jshi33/.)

Jianjun "Jan" Shi, PhD

Stephen Chick joined the IOE Department in 1995 as an associate professor after receiving his PhD in industrial engineering and operations research from the University of California–Berkeley. He worked for five years in the automotive and software industries before joining academia. His research brings together simulation and statistical decision-making tools to help improve process design decisions in a variety of industries, including the health sector. His research has been funded by the US Environmental Protection Agency, the Centers for Disease Control and Prevention, and the National Institutes of Health, and he has worked with several pharmaceutical, vaccine, and health care delivery organizations. He has served as president of the Institute for Operations Research and the Management Sciences' Simulation Society and is on the editorial boards of the journals *Management Science*, *Operations Research*, and *Production and Operations Management*. In 2003, he became professor of technology and operations management at the INSEAD school of business, which has campuses in France, Singapore, and Abu Dhabi, where he later held the Novartis Chair of Healthcare Management. (See Chick's page at the UM Faculty History Project: http://um2017.org/faculty-history/faculty/stephen-e-chick.)

Stephen E. Chick, PhD

Marina Epelman joined the IOE faculty in 1999 after completing her PhD in operations engineering from MIT. Her research has focused on the development and optimization of complex stochastic networks, with applications in scheduling manufacturing maintenance operations, distributed service systems, and cancer radiotherapy procedures. She has received awards for her research from INFORMS and for her teaching from Alpha Pi Mu. (See Epelman's curriculum vitae: www-personal.umich.edu/~mepelman/vita.pdf.)

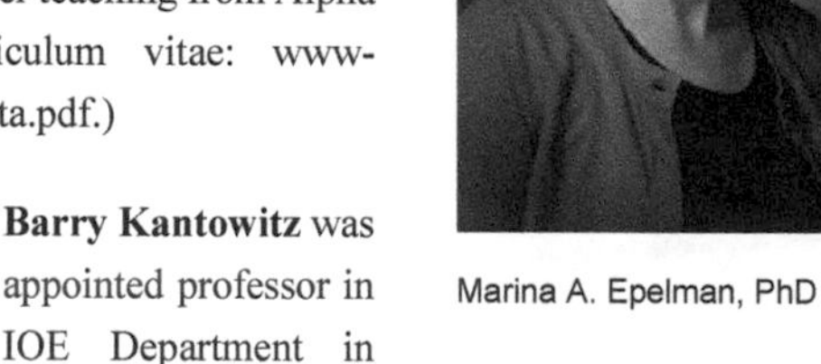

Marina A. Epelman, PhD

Barry H. Kantowitz, PhD

Barry Kantowitz was appointed professor in IOE Department in 1999 and director of the University's Transportation Research Institute, a position he held until 2004. In 1969, Kantowitz received his PhD in psychology from the University of Wisconsin. He joined the Purdue University faculty as an assistant professor in 1969, and rose to full professor in 1979. He served as chief scientist in the Human Factors and Organizational Effectiveness Research Center at the Battelle Memorial Institute (1987–1993) and as the human factors scientific adviser to the NASA Aviation Safety Reporting System (1991–1999). He was also director of the Battelle Human Factors Transportation Center from 1993 to 1995 and chief scientist from 1993 to 1999. Kantowitz's research focused on human attention and information processing. His applied ergonomics research has improved safety and efficiency in aviation, nuclear power, and intelligent surface transportation systems. He is the author of more than 100 papers in peer-reviewed journals, more than two dozen books, and 18 book chapters. He retired from IOE in 2012. (See Kantowitz's page at the UM Faculty History Project: http://um2017.org/faculty-history/faculty/barry-h-kantowitz)

Lawrence M. Seiford, PhD

Larry Seiford joined the IOE Department in 2000 as professor and chair. He received his PhD from the University of Texas–Austin in 1977. He has served on the faculty at the University of Massachusetts, the University of Texas at Austin, the University of Kansas, and York University. Seiford also served as program director of the Operations Research and Production Systems programs at the National Science Foundation from 1997 to 2000. Seiford's teaching and research interests are primarily in the areas of quality engineering, productivity analysis, process improvement, distributed-systems design issues, and performance measurement. In addition, he is recognized as an expert in the methodology of data envelopment analysis. Results of his research efforts have been incorporated in academic course offerings and practitioner workshops on benchmarking, productivity, and performance measurement. He has authored and coauthored four books and more than 100 articles in the areas of quality, productivity, operations management, process improvement, decision analysis, and decision support systems. He is a fellow of the Institute of Industrial Engineers (IIE), the American Society for Quality (ASQ), and the Institute for Operations Research and the Management Sciences (INFORMS). (See his UM web page: www-personal.umich.edu/~seiford.)

Jussi Keppo, DrTech

Jussi Keppo joined the IOE Department in 2001 after holding a postdoctoral position at Columbia University. He received his Dr.Technology in applied mathematics from the Helsinki University of Technology in 1998. His research focuses on stochastic control, statistical analysis of stochastic processes, and optimization methods, with applications in information economics, banking regulation, optimal investment under uncertainty, and risk management. He has had several publications in top-tier journals, such as the *Journal of Economic Theory*, *Review of Economic Studies*, and *Journal of Business* on such topics as investment analysis, information economics, and banking regulation. His research has been supported by several Asian, European, and US agencies, including the National Science Foundation. He left UM in 2012. (See Keppo's page at the National University of Singapore website: http://bizfaculty.nus.edu/faculty-profiles/314-jussi-keppo.)

Amy Cohn joined the IOE faculty in 2002 after receiving her PhD from the Operations Research Center at MIT. Cohn is as an affiliate of the MIT Global Airline Industry Program and is actively involved with INFORMS, AGIFORS, and the Industry Studies Association. Her primary research interest is in robust and integrated planning for large-scale systems, predominantly in health care and aviation applications. She also works on a number of other applied research projects, including collaborations in satellite communications and robust network design for power systems. Cohn is also known for her outstanding contributions to teaching engineers. Most recently she assumed the codirectorship of the Center for Healthcare Engineering and Patient Safety (See Cohn's UM web page: http://www-personal.umich.edu/~amycohn.)

Amy M. Cohn, PhD

Nadine B. Sarter, PhD

Nadine Sarter received her PhD in industrial and systems engineering, with a specialization in cognitive ergonomics/cognitive systems engineering, from Ohio State University in 1994. She joined the IOE faculty as an associate professor in 2004, after serving on the faculty at the University of Illinois and the Ohio State University. Sarter's primary research interests include multimodal interface design (with an emphasis on tactile feedback and crossmodal attention), attention and interruption management, decision support systems, adaptive function allocation, and human error/error management. She has conducted her work in a variety of application domains, most notably aviation and space, medicine, the military, and the automotive industry. For her work on pilot-automation interaction, she received the Aviation Week and Space Technology's Laurels Award for Outstanding Achievement in the Field of Commercial Air Transport in 1996, a National Science Foundation Faculty Early CAREER Award in 1998, and the Human Factors and Ergonomics Society's Ely Award for the best paper published in *Human Factors* in 2008. Sarter is associate editor for *Human Factors and Applied Ergonomics* and was a member of the National Research Council Transportation Research Board Committee on Electronic Vehicle Controls and Unintended Acceleration (2010–2012). She was also a member of the Federal Aviation Administration Performance-Based Operations Aviation Rulemaking Committee/Commercial Aviation Safety Team Flight Deck Automation Working Group (2006–2013) and served as an expert witness in the 2013 National Transportation Safety Board investigative hearing on the crash of Asiana Flight 214. In 2015 Sarter was appointed director of the Center for Ergonomics. (See the Human-Automation Interaction and Cognition Lab website: http://thinclab.engin.umich.edu.)

Judy Jin joined the IOE Department as an associate professor in 2005 after serving on the faculty in the Department of Systems and Industrial Engineering at the University of Arizona for five years. She received her PhD in the industrial and operations engineering at UM in 1999. Jin is a recognized leader in quality engineering. Her research has focused on data fusion and system informatics for better comprehension and operation of engineering systems and decision making for quality and reliability assurance. Her research innovation and broad industrial impacts have been recognized with numerous awards, including nine best paper awards from *IIE Transactions*, the Industrial Engineering Research Conference, ASME International Mechanical Engineering Congress, and International Conference on Frontiers of Design and Manufacturing; the Forging Achievement Award from the Forging Industry Educational and Research Foundation, and CAREER and PECASE Awards from the National Science Foundation. Recently, she has been serving as a departmental editor for *IIE Transactions* and an editorial board member for *Journal of Industrial and Production Engineering*. She also serves as the vice president of international activities of INFORMS; the chairperson of the Quality, Statistics, and Reliability division of INFORMS, and the president of the Quality Control and Reliability Engineering division of IIE. Jin is a professor in the IOE Department and the director of the Manufacturing Engineering Program of Integrative Systems and Design Systems. (See Jin's UM web page: http://www-personal.umich.edu/~jhjin.)

Jionghua (Judy) Jin, PhD

Mark Van Oyen joined the IOE Department in 2005 as an associate professor after serving on the faculties at Loyola University of Chicago and Northwestern University. He received his PhD in the electrical engineering systems from UM in 1992. His area of expertise emphasizes the development and use of stochastic methods to improve operations and systems, including the scheduling of production systems, flexible workforces and policies to support effective processes, and health care operational improvement. He has received two grants for teaching, as well as grants for his research from the National Science Foundation, the General Motors Foundation, and the Electrical Power Research Institute. Van Oyen has received best paper awards from Institute of Industrial Engineers, and was selected as an ALCOA Manufacturing Systems Faculty Fellow. His teaching in IOE has included the cre-

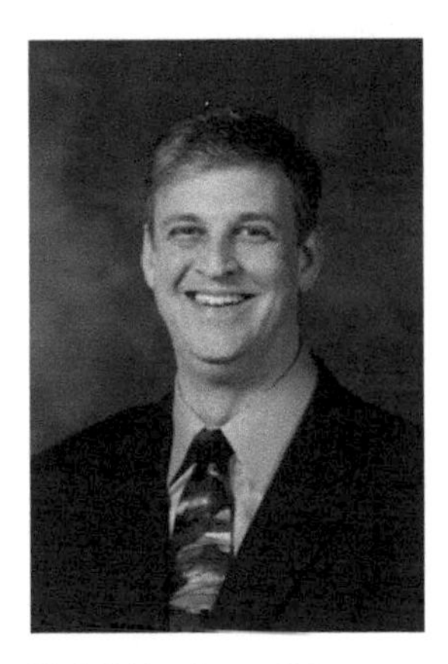

Mark P. Van Oyen, PhD

ation of several courses: IOE 440, Operations Analysis and Management (which included operations, supply chain, and queuing analysis of services); ENG 480, Global Synthesis Project, a revision of IOE 574, Simulation Analysis; a revision of IOE 545, Queueing Networks; and the senior capstone design course IOE 481, Practicum in Hospital Systems. He had been involved in the Engineering Global Leadership program for the College of Engineering in the roles of director, faculty adviser to the student honor society, and member of the admissions committee. During that time the program nearly doubled its enrollment and expanded from a program serving only IOE and mechanical engineering to one engaging students from almost every department of the College. He has served as an associate editor for *IIE Transactions*, *Naval Research Logistics*, and *Operations Research* and as senior editor for *Flexible Services and Manufacturing.* (See Van Oyen's UM web page: http://vanoyen.engin.umich.edu/.)

In addition to these faculty members, nine others joined the faculty for shorter periods: Medini Singh (1990–1993), Tava Olsen (1994–2001), Rachel Zhang (1994–2001), Shane Henderson (1996–2002), Mark Lewis (1999–2005), Sebastian Fixson (2002–2006), Sushyant Sharma (2002–2007), Goker Aydin (2003–2009), and Volodymr Babich (2003–2009).

Though the department was able to hire a number of outstanding people during this 15-year period, the following tenured faculty members retired or accepted academic positions at other institutions: Jack Lohman, Nandyan Srinivasan, Candace Yano, Walton Hancock, John Birge, Jim Miller, Dan Teichroew, Jim Bean, and Steve Pollock.

5.6. The 2000 Curriculum

The curriculum in 2000 was similar in structure to what was presented 10 years earlier, with the exception that students were given more flexibility in electing courses in non-IOE subjects, both within engineering and outside engineering. This flexibility has probably contributed to the popularity of the undergraduate program.

The master's degree had course requirements that continued to be flexible during this period. The master's degree requirement simply stipulated that students were to select about two-thirds of their courses from courses at the 400 level and above, and they could combine these with courses outside the department that were approved by a faculty adviser. This would allow students to obtain a professional-level knowledge in a subspecialty of interest, such as ergonomics, financial engineering, global leadership, or transportation systems.

IOE Undergraduate Curriculum in 2000

Course	Credit Hours
Subjects required by all programs (52 hours)	
Mathematics 115, 116, 215, and 216	16
Engr 100, Intro to Engr	4
Engr 101, Intro to Computers	4
Chemistry 125 and 130 (5)	4
1 hour applied according to individual program directives	
Physics 140 with lab 141; 240 with lab 241 (10)	8
2 hours applied according to individual program directives	
Humanities and Social Sciences	16
Related engineering subjects (12 hours)	
Non-IOE engineering courses	12
Required program subjects (28 hours)	
IOE 201: Industrial Operations Modeling	2
IOE 202: Operations Modeling	2
IOE 310: Intro to Optim Methods	4
IOE 265: Engr Probability and Statistics	4
IOE 333: Ergonomics	3
IOE 334: Ergonomics Labs	1
IOE 316: Intro to Markov Processes	2
IOE 366: Linear Statistical Models	2
IOE 373: Data Processing	4
IOE Senior Design Course	
[IOE 424 (4) or 481 (4) or 499 (3)]	(3)4
Technical electives (24 hours)	24
Unrestricted electives (12 hours)	12
Total	128 hours

The PhD program course structure continued to emphasize six areas: applied statistics, engineering economy and management engineering, ergonomics, information systems, operations research, and production and manufacturing. This broad selection of courses also contributed to the popularity of the IOE PhD degree during this period, and more than 75 PhD students were enrolled in 2001.

A typical PhD program resembled the following. During the first year the students were required to complete at least 24 credit hours of mostly IOE courses. After successfully completing these courses with high grades, they would then take a written qualifying examination, based on courses they had taken in three of the six areas of concentration. They would also propose a research topic. If they passed this exam, they would commence their initial research. After completing their second year, they would write a thesis proposal and present it to a prospective thesis committee, one member of which had to be from outside the department. The thesis committee would then conduct a preliminary examination of the thesis topic. This examination included the written proposal and an oral presentation describing the proposed research. It allowed the student and faculty members to fully understand the intent of the research, its scope, its potential impact, and the resources required to perform the research. Successful completion of the preliminary examination allowed the student to become a candidate for the PhD degree and commence their thesis research.

Securing the financial and other resources necessary to complete the required PhD thesis research could be a daunting task for a student. Most often it depended on the financial support provided by the primary faculty thesis advisers at the time of the preliminary examination. Since a PhD thesis research project could cost over $250,000, particularly in areas that required extensive experimentation and/or fieldwork, the primary thesis adviser and student had to form an effective team to secure funding. Often the sources and types of outside grants and contracts secured by IOE faculty members shaped the nature and ultimate quality of a PhD thesis. This financial reality required successful faculty members to aggressively seek and secure the type of funding needed to ensure that high-quality PhD thesis research could be performed. Assuming that adequate resources could be provided, the PhD thesis research would typically be completed two or three years after the thesis preliminary examination. To conclude the PhD program, the student would submit a written doctoral thesis to the thesis committee and stand for an oral, public presentation and examination of their research by his or her committee.

5.7. Funding Sources

As mentioned earlier, the IOE PhD program had become very popular during the ’90s, but this also created a large financial burden on the faculty. It also should be kept in mind that at this same time the undergraduate and master’s degree programs were demanding significant time and effort from the faculty. To their credit, the faculty members managed to find support for their own research and that of their students from a variety of sources. The 2003 departmental review indicates that the faculty had secured more than $15 million in sponsored research in the previous five years. Of this, the federal government provided about $10 million; about $3 million came from industrial groups; and the rest came from foundations, states, and other sources. As would be expected because of the highly empirical nature of the research related to ergonomics and manufacturing quality engineering, almost $10 million was expended in these two areas of concentration, followed by about $3 million in operations research, $1.8 million in production systems, and $1.4 million in engineering management and financial engineering.

5.8. Staff Support

In addition to the technical staff support described earlier, as the size of the IOE faculty and student body increased from 1990 to 2005, and government reporting requirements became more demanding, the department needed a competent staff to provide administrative support. Some of the people who have provided continuing and important assistance over the past couple of decades are shown here.

Candy Ellis – IOE Departmental Administrator

Elizabeth Fisher – Marketing and Communications Administrative Assistant

Kelly Cormier – Center for Ergonomics

Tina Picano Sroka – Department Assistant

Mary Winter – Research Administrator

Wanda Dobberstein – Undergraduate Student Adviser

Matt Irelan – Graduate Student Adviser

5.9. Synopsis of 1990–2005—IE Contributions and Major Events

Many significant and important events occurred during this 15-year period. One of the most prominent was the completion of the expanded IOE building in 1996. This provided lecture and seminar rooms as well as new teaching, research laboratory, and shop spaces. But it wasn't only the physical facilities that were improved; the hiring of additional information technology support staff allowed faculty and students to use and even develop new computer programs to address many contemporary and important problems in the field.

This 15-year period also saw a significant (about 40 percent) increase in enrollments, particularly at the undergraduate and master's levels. It was also a period in which female students now made up about 40 percent of the undergraduates, 25 percent of the master's degree students, and 22 percent of the PhD students. Thirteen percent of the undergraduates, 11 percent of the master's students, and four percent of the PhD students were classified as being from underrepresented minority populations from 1997 to 2002. It was also during this period that the faculty had the largest number of women faculty members.

The tuition provided by the additional students electing IOE degrees supported the hiring of many new faculty members, resulting in the highest number of tenure-track faculty members (25) in the department's history. There was also a threefold increase in the number of adjunct faculty members hired during this period to assist in teaching the lower-level courses. One of the benefits of the larger number of faculty members was that a critical intellectual mass in ergonomics, operations research, production systems, quality and manufacturing engineering, and management and financial engineering existed for most of this period. There was also a good balance between theoretical and applied research.

During this period several of the senior faculty members in the department developed exciting new educational programs in cooperation with faculty members from other UM schools. Their leadership resulted in the Tauber Manufacturing Institute, a joint venture with the School of Business, the first College of Engineering honors program in engineering global leadership, and a popular financial engineering master's Program that involved five different schools and colleges. The structure and impact of these new programs, along with other such continuing initiatives lead by IOE faculty members is described in the next two chapters.

There was also an increase in the number of well-qualified PhD students in the department during this period, reaching 74 in 2004. Though this was a positive turn of events, it created an increased need for the 25 faculty members to find support for the research projects these students wished to perform. Although the faculty members were able to increase their sponsored research, and by 2001 had secured an average of over $200,000 per faculty member (higher than other peer schools), the 2003 departmental review indicated that most faculty members believed this was still not sufficient for the size of the PhD program. The pressure to find increased funds to support students while teaching courses with a large number of students resulted in the loss of some excellent faculty during this period.

Alumni relations were also greatly improved during this period with the publication of a much more comprehensive biannual *IOE Newsletter* and the establishment of an IOE Alumni Academy that met several times each year to discuss how alumni could assist the department. An Outstanding Alumni Award was established, along with the endowed Bert Steffy Lectureship Awards (See Appendix A.1 for a description of the winners of these awards.)

In summary, during this 15-year period the IOE Department had grown in size, impact in its field, and prestige. As part of one of the preeminent, large public universities in the United States, with more than 120 different departments, it is clear that after 50 years the IOE Department was now doing its part by providing outstanding industrial engineering leaders in many different areas of endeavor.

[6]

Building Centers and Programs

Six important collaborative efforts were developed through the leadership of industrial and operations engineering (IOE) faculty members over the past 50 years:

- Information Systems Development and Optimization Systems (ISDOS)
- The Center for Ergonomics
- The Dynamic Systems Optimization Laboratory
- The Engineering Global Leadership Program
- The Tauber Manufacturing Institute
- The Financial Engineering Master's Program
- The Hospital Systems Engineering Program

Each of these programs had unique qualities and had a significant impact on the education and research within the department. In many ways the department's continued high national ranking can be attributed to the leadership necessary to develop and operate these collaborative programs. In the following sections each of these programs is described.

6.1. Establishing a Major Industrial Engineering–Oriented Information Systems Organization—ISDOS

6.1.1. Background of ISDOS

As described earlier, in the '60s the department relied on Richard Wilson to develop and instruct students on how to use digital computers to create analytical operations research models. In addition, Dean Wilson drew upon his expertise in the iconic American Airline's Sabre reservation system to teach the use of various types of programmable digital computers in manufacturing and signal processing, as well as in implementing the Monte Carlo system-simulation method. Bert Herzog, who had joined the department in 1965, was directing research and teaching in computer graphics and networks. In 1966, these three were joined by Ed Sibley from MIT, whose expertise was in management information systems. Dan Teichroew had begun his work on Problem Statement Language and its supporting software, Problem Statement Analyzer at Case Western Reserve in the early '60s. In 1968, Teichroew joined the department as its chair. Soon after, in 1970, Alan Merten was recruited from the University of Wisconsin's Computer Science Department.

Thus, by 1970, six faculty members were instructing in and researching the use of digital computers and computer language development to model and analyze complex information technology systems.

Dan Teichroew in the ISDOS Library (1983).

Dean Wilson, an early information technology leader (1965).

Alan Merten and Dan Teichroew at a social event in the '90s.

6.1.2. Overall Focus of Research in ISDOS

Teichroew's concern in the late '60s was that information systems lacked the equivalent of a manufacturing blueprint—a coherent, complete plan of what is going to be built that ensures that the construction will meet all the requirements. So the initial thrust of the ISDOS project was to develop a mechanism for describing an information system and its requirements. The mechanism was linguistic because, in the 1960s, computer graphics were just starting to become available. In addition, personal computers had not yet been developed as local workstations, so all development was on a mainframe computer.

The blueprint modeling mechanism that the ISDOS team developed, initially at Case Western Reserve University in 1967, was called the Problem Statement Language (PSL). PSL modeling was based on an extended entity-relationship model. PSL statements describing an information system and its requirements were presented to a repository managed by the Problem Statement Analyzer (PSA) software. PSL statements were synthesized as they were presented and only checked for syntactical correctness and lack of contradiction at the time of entry, but not for completeness.

The PSL statements could be added into the model by different analysts at various times, so a system could be described in modules or pieces (similar to the concept of modular, agile programming). Then, when the PSA analyzed all the pieces, the model could be refined and completed by discovering gaps, resolving design issues, identifying overlaps, and so on. PSA could generate reports on different aspects of a system model from the contents of the repository and check for model completeness. When the system description was deemed complete, it could be invoked to produce a composite description and specifications for the information system under development.

6.1.3. Innovative Strategy for Funding ISDOS Research

Teichroew proposed that ISDOS should be funded by contributions from potential users, he termed them "sponsors." who paid a modest yearly subscription fee. In return, the sponsors could attend yearly workshops at Michigan and selected off-site locations. Sponsors were given the research products to take back with them and apply to their work. (This necessitated the implementation of PSA software across various environments, such as IBM, Honeywell, Burroughs, CDC, and DEC.) Sponsors were asked for feedback on the utility of the ISDOS development products. This resulted in unheralded techniques of storing the software in a database and mapping it into various environments. Some sponsors wanted PSA reports to match the framework of their documentation standards, and the resulting ISDOS product was the Automated Documentation System, which formatted PSA reports to an organization's standardized content.

ISDOS products were developed by IOE graduate students (five to 10 at a time) guided by faculty and senior researchers. Results were published in books and national journals. For example, the key article "The Automation of Systems Building" was published in *Datamation* in 1970 and later included in the books *Systems Analysis Techniques* (edited by J.D. Cougar and R. Knapp, John Wiley & Sons, 1974), and in *Classics in Software Engineering* (edited by Edward Nash and Ed Yourdon, Prentice Hall, 1979). Yourdon and Teichroew developed a relationship that resulted in PSL being used to formalize Yourdon's hitherto informal Data-Flow Diagrams. Worldwide exposure from these publications stimulated interest in ISDOS work from foreign organizations. Teichroew was invited to make presentations about his work and consequently drew foreign sponsorships for ISDOS.

The sponsorship model may look very familiar to today's researchers, but it was innovative and unusual at the time. It had the beneficial side-effect of mitigating conflict-of-interest issues between researchers and the people who funded them and reducing dependency on one or two major sponsors. Sponsors who did not have or did not want to spend a substantial amount could still benefit from the innovative research and development products. Essentially, Teichroew had established and was operating one of the first consortiums in the University, and at its peak in 1982 ISDOS was providing more than $1.1 million in sponsored research funds to the IOE Department (the equivalent of about $2.6 million in 2014).

6.1.4. Major Users and Supporters of ISDOS

United States sponsors included the US government, including many Department of Defense organizations and contractors as well as commercial organizations. The Social Security

Administration was a major user of PSL/PSA and required all of its contractors to use PSL to describe the information systems they were proposing to develop.

Military users included the US Navy. Groups such as the Naval Ocean Systems Command, Naval Underwater Systems Command, Naval Ship Research and Development Center, Naval Surface Warfare Center, and Naval Weapons Center at China Lake used PSL/PSA. When the Navy wanted to procure a computer-aided laboratory, it developed the request for proposal, that is, its requirements, using PSL/PSA and the Automated Documentation System. Various groups in the US Air Force also sponsored the research. A major request for proposal released by the Air Force was in the form of PSL statements. A special version of PSL/PSA was developed for the Electronic Systems Division at Hanscom Air Force Base in Massachusetts called URL/URA under the MULTICS operating system. Corresponding courses were also designed.

In the corporate arena, Raytheon, Rockwell International, Computer Sciences Corporation (CSC), and TRW were among contractors using PSL/PSA; TRW developed a specialized version of PSL/PSA that they labeled RSL/REV under an umbrella called SREM. IBM obtained a corporate-wide license for PSL/PSA. Other commercial users included the Boeing Corporation, National Cash Register (NCR), the Upjohn Company, Mobil Oil, Diners Club, Pacific Gas and Electric in Southern California, Honeywell, Burroughs, CDS, DEC, and computer-aided software engineering tool vendors, such as Index Technology of Cambridge and the NASTEC Corporation of Southfield, Michigan. Various banks also used PSL/PSA. Among the foreign users were the August Thyssen Group in Germany, Dutch Post Telephone & Telegraph, British Rail, ARAMCO, and L. M. Ericsson.

The ISDOS/PRISE Computer Room in the IOE Building (1984).

6.1.5. ISDOS Becomes a Commercial Spin-off

To help showcase to the taxpayers of Michigan how university research contributed to the state's economy, the ISDOS project was selected and spun off (around 1983) as ISDOS Inc., headquartered in Ann Arbor. Elliot Chikofsky was chief of marketing. ISDOS Inc. was eventually bought by Methods Works Inc. of Canada. It, in turn, was sold to a British company. Richard Welke, then principal of Methods Works went on to Georgia State University, where he was appointed professor of computer information systems and director of the Center for Process Innovation.

During the transition years to a commercial company (1983–1985), much of the research work was left in the IOE Department. To avoid confusion, in 1983 the department laboratory was renamed the Program for Research in Information Systems Engineering (PRISE). Some of the sponsors of ISDOS continued to support Teichroew's efforts in PRISE, which included supporting several PhD students. Unfortunately, at the time the University General Counsel's Office did not have a clear policy on how a sponsor could divide funds between the UM's PRISE Laboratory and ISDOS Inc. without creating a conflict of interest for Teichroew. The result was that PRISE was not well funded after 1987. Also, as computer science departments emerged in various academic institutions, including UM, the research work on high-level program languages shifted away from industrial engineering (IE). By the early '90s the PRISE Laboratory had closed.

6.1.6. Influence on ISDOS Graduates

ISDOS influenced many graduate students. Jay Nunamaker, a student of Teichroew's who worked on an early version of PSA, went on to become dean of the College of Business and Public Administration at the University of Arizona. One of the 1973 IOE PhD graduates, Hasan Sayani (whose major adviser was Ed Sibley) formed an informal "ISDOS East" at the University of Maryland. With help of a cognitive psychologist and curriculum designer, Cyril Svoboda, PSL/PSA courses were revamped. Special courses meant for the larger population of PSA report "readers" were designed, along with corresponding, more detailed courses, for the "writers."

Sayani was also instrumental in having PSA implemented on the UNIVAC computer system at the University of Maryland; this was then implemented on the Social Security Administration's computer system. Sayani implemented PSA on the HP/UX systems and sold it as a turnkey system (e.g., to the Naval Surface Warfare Center). Several associated systems were developed by the ISDOS East group as well; among them, a reverse engineering system (REVENGG), a logical database design system, and the Automated Documentation System, which incorporated standard text with PSA reports. The Automated Documentation System dynamically inserted PSA reports within free-form text. Two-way bridges were developed between PSA and computer-aided software engineering tool products.

Other IOE PhD graduates who were supported and guided by Dan Teichroew, Ed Sibley, Alan Merten, and Tony Woo during the '70s and '80s were: William Ash ('71); Milton Drott ('73); David Carlson ('75); Koichi Yamaguchi ('75); S. Navathe ('76), who was named an Association for Computing Machinery fellow in 2014; David Johnson ('81); Kyo Kang ('82); Timothy Thomasma ('83); Sung Yong Shin ('86); and Hyun Lee ('88). All went on to have prominent positions in the information technology world.

6.1.7. ISDOS Work Overtaken

Although many sponsors and clients found PSL/PSA an exciting advance during the '70s and '80s, attempts at commercialization were ultimately stymied. Though PSL/PSA principles were integrated into computer-aided software engineering tools, capabilities were overpromised, eventually discouraging sales. A common organizational issue also arose with management impatience with the rigor and analyst time required to produce good information technology requirements and specifications. And though Dan Teichroew continued his work within the IOE Department until his retirement in 2001, Hasan Sayani invited Ed Sibley to join the Information Systems Management Department at Maryland in 1973. When that was disbanded at College Park, Sibley left for the Department of Information and Software Systems Engineering at George Mason University, where he is now a professor and eminent scholar. Sayani went on to direct the graduate program in software engineering at University of Maryland University College, where he drew upon his prior experience to develop a commercial Semantic Database Management System—CaMERA. In 1975, Bert Herzog left UM to join the engineering and computer graphics faculty at the University of Colorado. He later returned to UM in 1987 as director of the Computing Center. Alan Merten transferred his appointment to the UM School of Business Administration in 1976, and after appointments at Cornell University and University of Florida, he went on to become president of George Mason University in Virginia (he recently retired). Tony Woo, who joined the IOE Department in 1977 and was instrumental in the development of models and computer algorithms to guide the geometric assembly of complex shapes, left in 1990 to join the IE department at the University of Washington.

6.2. The Center for Ergonomics

In 1979, the ergonomics faculty within the IOE Department requested that a designated center for research on ergonomics issues be established within the College of Engineering. After review by the College administration and the vice president for research's advisory board, in November 1979 the regents of the University granted the request, and the Center for Ergonomics was created. Because of its scope of activities and its importance for more than 35 years, chapter 7 is devoted to describing the history of the center and its impact on the field of ergonomics.

6.3. The Dynamic Systems Optimization Laboratory

The Dynamic Systems Optimization Laboratory was formed in 1985 and is housed today in a computer lab on the main floor of the IOE Building. It was created by Professors Robert Smith, James Bean, and Jack Lohmann to serve as an applied research consortium that could unite the shared interests and activities of those faculty and students engaged in the modeling and analysis of problems involving dynamic, sequential decision making over time. The lab was novel in its establishment of a common space with shared computational facilities for operations research faculty and their students to engage with real-world industrial firms and governmental entities to tackle some of the most demanding problems confronting the world. Its primary emphasis was on improving the design and operations of complex systems. The lab's original computational facilities were provided by a special grant from Bell Laboratories that specifically supported the lab's work on establishing the existence and determination of planning horizons for long-term strategic planning. The National Science Foundation and General Motors provided many subsequent grants for this field of infinite horizon optimization, totaling roughly $1 million over the years. The PhD students at the lab whose dissertations were written in the arena of infinite horizon optimization and asset planning included David VanderVeen, who later became director of product development analytics at GM; Matt Bailey, who acquired the title of Howard J. Scott Professor of Management at Bucknell University; and Sarah Ryan, who became a professor at Iowa State University and an Institute of Industrial Engineers fellow. Also the National Science Foundation CAREER Awardee Allise Wachs, who later became president of Integral Concepts, and Tim Lortz, who is on the technical staff at Booz Allen, were early contributors. Two other notable PhD students whose dissertation research in infinite horizon optimization formed some of the theoretical foundations for the laboratory were Wally Hopp, who became the Herrick Professor of Business and senior associate dean in the business school at the University of Michigan, and Julie Higle, who later became chair of industrial and systems engineering at the University of Southern California.

Robert Smith, PhD

James Bean, PhD

Jack Lohmann, PhD

One of the other recurring themes of the lab was the challenge of designing and managing the flow of vehicular traffic to relieve delays caused by recurring and incident-caused traffic congestion. The lab tackled this problem in the private and governmental sectors through dynamic route guidance and coordinated traffic-signal control. Stephane Lafortune from the EECS Department played a prominent role on the problem of coordinated signal control. This research was supported by the FHWA Intelligent Transportation Systems Research

Center of Excellence at the UM. Participants included PhD students Karl Wunderlich, who later became a tech fellow at Globis Corp, and David Kaufman, formerly at AT&T Labs. The approach was to animate the vehicles and traffic signals within a very large game played by tens of thousands of players within a computer simulation that discovered the best cooperative strategies though a fictitious-play algorithm.

This fictitious-play algorithm was later generalized with the support of several National Science Foundation grants to orchestrate the design and operation of general complex systems. One application of fictitious play was the design of a dedicated production line, supported by grants from General Motors through the GM Collaborative Research Lab at UM. This application integrated revenue management and inventory control through their interplay within fictitious play. As PhD students, Archis Ghate, associate professor at the University of Washington, and Shih-Fen Cheng, associate professor at Singapore Management University, played central roles in this research. Another application, supported by a Multidisciplinary University Research Initiative (MURI) grant from ONR, provided an algorithm for minimum-time path finding for naval vessels in an evolving ocean wave field. Another MURI hrant, this time from the Army Research Office, for the design of low-energy mobile communication electronics for situation awareness, led to implementation of the hit-and-run algorithm developed by Smith within a large-scale simulation optimization algorithm for overall design of the mobile electronics system. Hit-and-run was later incorporated by Zelda Zabinsky and one of her PhD students working with Boeing Corporation, which resulted in a software suite for the design of composite airplane components. This software was later used in the design of the Boeing Dreamliner. Hit-and-run is often regarded as the most efficient platform for Monte Carlo generation within complex stochastic simulation samplers.

The lab's funding for these activities was large scale for the field of operations research, numbering in the hundreds of thousands of dollars per year. Other PhD students who served as research assistants within the lab and worked in several of the research areas were Jeff Alden and Dan Reaume, who later became tech fellows at the General Motors Research Labs; Edwin Romeijn, who joined the IOE faculty for several years and later assumed the chairmanship of the School of Industrial Engineering at Georgia Tech; Alfredo Garcia, who became professor of industrial and systems engineering at the University of Virginia; and Irina Dolinskaya, who later held the William A. Patterson Junior Chair in Transportation at Northwestern University. Jeff Alden later won INFORMS' most prestigious prize, the 2005 Franz Edelman Award for Achievement in Operations Research and the Management Sciences.

6.4. The Engineering Global Leadership Program

In 1992, IOE professor James Bean established the Engineering Global Leadership (EGL) program, a five-year honors program designed to strengthen the IOE educational program by addressing three recognized issues: (1) the need for young engineers (particularly new degree recipients) to develop and improve their communication skills to breach the engineering-business boundary, (2) the increasing requirement that graduates learn how to comfortably and effectively employ their skills and abilities when working within another culture, and (3) correct the lack of a formal honors program within IOE (and in the College of Engineering) that could recognize and promote individual excellence and global leadership skills.

The core curriculum of the EGL program addressed the first two of these gaps by combining the then traditional engineering curriculum with core courses in the School of Business Administration; the School of Literature, Science, and Arts; and selected foreign educational institutions. The former explicitly addressed leadership and organizational issues, and the latter exposed students to the languages, history, and customs of a student-selected non-US country or region of the world with existing or potential competitive and engineering presence.

The curriculum also required students to complete a synthesis team project that placed them within a specific and real industrial context, to which they had to apply their technical knowledge and develop their teamwork skills. This exposure of excellent students to organizations with operational and other engineering missions, and (perhaps more important) exposure of the organizations to these students, was an immediate success.

The EGL program was also an honors program, since the admissions (and subsequent academic performance) requirements were very high. Even though its added course requirements and constraints were time consuming, the benefits were clear, and from the very beginning the enthusiasm and accomplishments of the students were exemplary.

At first, the program led to two degrees after an average of five years at Michigan: a BSE and then an MSE in industrial and operations engineering. By the year 2000 the EGL program was made available to a small number of students in mechanical engineering and manufacturing engineering. (Over the following five years, under the guidance of then-EGL director Stephen Pollock, the program was further extended to students in all other College of Engineering degree programs, although the majority of EGL students still obtained their degrees in IOE.) In the mid-1990s, enrollment in the EGL program was five to 12 students per year, but by the early 2000s, after the incorporation of non-IOE students, EGL enrollments grew appreciably and then leveled off at about 50 per year.

The payoff of this innovative program was immediate: employers recognized that EGL students were not typical master's degree graduates, and that they did not fit into the traditional engineering job role. In response, some companies actually created new positions and job rotation programs specifically for EGL graduate, to take advantage of their unique skills and talents. By 2005, similar programs were instituted in other universities, but UM's IOE led the way with the first one that combined cross-disciplinary engineering-business education, requirements for understanding international cultural aspects of engineering, and honors recognition for academic and applied engineering excellence.

In the early days of the EGL program, student projects were sponsored by a wide variety of firms, including Allied Signal, Cummins Engine Company, DaimlerChrysler Corporation, Dell Computer Corporation, Ford Motor Corporation, Intel Corporation, and Lucent Technologies, as well as consulting firms such as A.T. Kearney, Bain & Company, Boston Consulting Group, Diamond Technology Partners, McKinsey & Company, and PricewaterhouseCoopers. By the turn of the century, student projects became integrated with the Tauber Manufacturing Institute (discussed in the next section), which provided a richer and more stable source of student projects sponsored by over 30 organizations each year.

6.5. The Tauber Manufacturing Institute

6.5.1. Early Years (1993–1999)

The concept for a multidisciplinary business and engineering enterprise at the University of Michigan was born in November 1991 as a result of a joint meeting between the College of Engineering and School of Business Advisory Boards. A Program Development Advisory Board—consisting of industry executives from 27 corporations—was assembled and held its first meeting in the spring of 1992. Through the course of many more meetings, the goals, core curriculum, and organizational structure of a new program named the Michigan Joint Manufacturing Initiative (MJMI) was launched in 1993.

IOE professor James C. Bean was appointed the codirector as part of his role in the College to broaden the scope of the undergraduate engineering curriculum beyond predominantly technical subjects. His leadership from 1993 to 1999 allowed the MJMI at the University of Michigan to be the equal of MIT's Leaders for Manufacturing. Both programs were created at a time when Japan and other overseas rivals were challenging US manufacturing dominance in many areas, including the automotive industry. Peter Banks, dean of the College of Engineering, and B. Joseph White, dean of the Stephen M. Ross School of Business, were at the helm when the innovative joint institute took shape.

Business school professor Brian Talbot and IOE professor Chip White were asked to join Jim Bean in leading the program. They recruited Craig Marks, who had just retired from TRW. Marks served as the institute's industry codirector from 1993 to 1999.

It is worth noting that the concept of such a joint educational program between two large and highly ranked schools was controversial at the beginning, but the success of the student team projects (described later) played a major role in the institute's gaining acceptance. Students were already aware that a cross-disciplinary education was the future, and once they heard about the program's details, they began gravitating toward the institute. US companies that were losing market share to overseas interests and other manufacturers were ready to invest in the creation of a new kind of employee—one with training in business and engineering.

A pivotal point in the existence of the MJMI came in 1993 when Peter Banks, who had strongly supported the institute, announced that he was resigning as dean of engineering. Some opponents of the institute were ready to let it dissolve at that point, but Joel Tauber, a UM alumnus and manufacturing executive from Detroit, gave the institute its second wind in the form of an endowment, and this led to a name change: MJMI became the Joel D. Tauber Manufacturing Institute (TMI).

TMI soon became one of the top multidisciplinary programs in the country. It was the only major manufacturing program that offered a full range of degree options for master's and undergraduate students. In addition to normal degree requirements, TMI students completed an intensive leadership program in engineering and business that culminated in a team-based, 14-week paid internship with a high-profile manufacturing company.

Team projects were (and continue to be) at the heart of the institute's philosophy, which focuses on teamwork, integration, and manufacturing success. Teams usually comprised two or three students from the School of Business and the College of Engineering, and the student team projects were designed to test students' collaborative skills. Working as engineers and business managers, teams collaborated to analyze a current manufacturing problem and design and implement a solution that would improve the sponsoring company's performance.

Though traditional manufacturing companies were key players in the early days of TMI, companies in electronics, pharmaceuticals, paint, and other industries came on board. In fact, Dell Corporation went on to hire a substantial number of Tauber graduates to work on its new supply chain strategies. Because Craig Marks was well connected with senior industrial leaders and knew that corporate America was looking for a new type of executive, he was able to provide TMI the business resources it needed for a number of years. He put together a team of industry people who were willing to work to help UM create a program that would shape the education of their potential new executives.

The open cooperation of industry executives created ownership of the new curriculum structure and resulted in additional financial support from the participating companies. The proposed team-project concept gained popularity, and many companies were anxious to support the students, who were soon competing for the most challenging projects. Word of the financial impact derived from the student project recommendations enhanced the competition among sponsoring companies to stay involved.

In 1999, Jim Bean left TMI to serve in two associate dean positions in the College of Engineering. Then, in 2004, he left UM for a position as dean at the University of Oregon, where he helped implement a similar team-focused program in their Sports Product Management Program. Today, Bean is provost and senior vice president for academic affairs at Northeastern University in Boston.

6.5.2. Later History (1999–2005)

Replacing Bean as the TMI engineering codirector was IOE professor Yavuz Bozer, who served the institute in that position until the fall of 2010. Bozer faced many challenges when taking this position. The very definition of manufacturing was changing, and some were claiming that manufacturing was dead in the United States. Labor-intensive work continued to be shipped overseas, and many plants in the United States were closing. Potential students were sensitive to these facts and expressed apprehension over joining a program with the word "manufacturing" in its very title.

Internally, the TMI codirectors worked with the admissions departments in the College of Engineering and the School of Business to inform students about the opportunities the institute presented. In order to continue on a successful path, The institute leadership doubled their efforts with the UM leadership to keep the institute's programs viable and attractive as an option for the best and brightest students. At this time the public interest in supply chain management, operations management, and manufacturing engineering in general began to rise.

One issue was to achieve agreement by the faculty and industry sponsors that a change was needed in the institute's name. Agreeing on the new focus and name was a daunting task. Multiple stakeholders were invited to suggest a new name. Alumni, prospective and current students, advisory boards, deans and faculty were all consulted. Finally, a new name was coined: Tauber Institute for Global Leadership (TIGL). (One year after the institute changed its name to reflect a global focus, MIT's Leaders for Manufacturing became Leaders for Global Operations.)

Project sponsors continued to evolve, and the team projects gained additional recognition as valuable endeavors for companies. To further the teams' contributions and learning and to smooth out any potential conflicts, a "team doctor" concept was initiated under Yavuz Bozer, which meant that in the rare event it was necessary, a staff member of the School of Business was deployed to help a team resolve occasional issues.

Bozer said that a well-known UM alumnus encapsulated Joel Tauber's impact in one sentence: "As far as I know, TIGL makes the best use of its money, and I am very impressed with the remarkable impact you made with a relatively modest donation." Funding, vision, and commitment have made TIGL an outstanding educational program for the College of Engineering and the School of Business. (For more information see the institute's website: www.tauber.umich.edu/.)

6.6. The Financial Engineering Master's Degree Program

The interdisciplinary Financial Engineering Program, offering a master's of science in engineering, was created in July 1996 under the leadership of John R. Birge, who was then the chair of IOE. The original participating departments were IOE, from the College of Engineering, and mathematics and statistics, from the College of Literature, Science, and the Arts. Although not part of the program initially, the Ross School of Business allowed financial engineering students to enroll in the required business courses, and in May 2002 it became a full partner in the program. By 2005, other participating departments included EECS (from the College of Engineering), economics (College of Literature, Science, and the Arts), finance (Ross School of Business), and the multidisciplinary Center for Complex Systems.

Birge became the first director of the Financial Engineering Program, and IOE participating faculty included Vadim Linetsky and Jussi Keppo. After Birge left for Northwestern, subsequent directors from IOE included Romesh Saigal, Larry Seiford, and Stephen Pollock.

The program started with an initial enrollment of five students, but by the fall of 2005, it had an entering class of around 60 students and had grown from a 30-credit program that could be completed in three semesters to a 36-credit program requiring a summer and three semesters. Many financial engineering students co-registered in PhD programs in such departments as economics, mathematics, and finance. Most had previous experiences in the financial industry.

By 2005, graduates had been hired by investment banks (e.g., Morgan Stanley), corporate treasuries (e.g., Ford Credit), financial consulting firms (e.g., Bain), software providers (e.g., Infinity), energy companies, (e.g., Detroit Edison), banks (Credit Suisse), mutual funds (T. Rowe Price), hedge funds (Susquehanna Partners), government (World Bank), and consulting companies (Author Anderson).

One mechanical engineering/IOE student in the program won the Eurobanco Commercial Bank competition by increasing a paper portfolio from $250,000 to $12 million in three months.

6.7. The Hospital Systems Research Program

In the early 1960s, the department increased its interest in improving hospital operations. This effort was lead by Clyde Johnson and Dean Wilson, the latter of whom was also serving as the director of the Industrial Systems Laboratory within the department. Johnson and Wilson collaborated with the University of Michigan Hospital to establish the first IE group within a hospital to improve its operations. Before long this approach had spread to other hospitals in Michigan, Indiana, and Ohio. Johnson provided the strategic guidance and coordinated the various projects. Wilson supervised the students in determining and documenting the time it took to perform many of the repetitive tasks that occur in hospitals. The intent was to provide a more precise method of determining the staffing levels needed in different departments and under various operating conditions. Numerous time studies were performed. Essentially, this massive effort was analogous to the way a manufacturing operation plans and determines its productivity goals.

Many IE students worked on projects in these hospitals, and full-time IE staff positions became available to IE alumni. At first many of the projects focused on reducing cost and waste in support functions, such as the laundry, housekeeping, and food services departments. By 1963, the demand for IE projects was sufficient to spin off a nonprofit company: the Community Systems Foundation. Wilson, along with a recent IE PhD graduate, Bart Burkhalter, and a recent IE undergraduate, Karl Bartsch, were the initial officers. Within a short period they had consulting projects in almost all the local hospitals and were providing a large number of IE undergraduates and master's students with excellent problems and experiences. The historical development of the foundation is described at its website: http://www.communitysystemsfoundation.org/downloads/CSF_50_Year_History.pdf.

During this period both Wilson and Johnson continued to teach in the IE Department, and in 1965, Richard Jelinek joined them as an assistant professor. Jelinek had just completed his PhD with Johnson as his adviser. Karl Bartsch left Community Systems Foundation and returned to the department to complete his master's degree. He then formed Chi Systems, a second spin-off, which provided additional hospital consulting services. Wilson resigned his faculty position in 1968 to devote himself to the rapidly growing Community Systems Foundation. About this time the department's Industrial Systems Laboratory, which had been providing staff support for the faculty and students to work in area hospitals, shut down because of political pressure from the now established private consulting groups, some of which were employing graduates of the department. In 1971, Jelinek moved to the faculty in the Department of Hospital Administration within the UM School of Public Health. Several years later he resigned from his faculty position to develop a new hospital systems consulting company in California. In 1974, Clyde Johnson retired, but his vision—along with his management of the many hospital projects performed by IE students and staff over his17 years with the department—had left a mark. The opportunities and means for all hospitals to improve their operations were now a matter of record, and companies like Community Systems Foundation and Chi Systems continued to expand their services throughout the '70s and '80s based on the earlier and continuing research being done in the department.

In 1970, Walton Hancock, who was chair of the IE Department and director of the Human Performance Laboratory, when studying ways to minimize hospital costs, became interested in whether the flow of patients was optimal. In order to learn more about patient flows and hospital management, he moved his primary office to the School of Public Health, where there was a hospital administration graduate program. There he joined John Griffin, the director of the program, and others in the School of Public Health to develop stochastic flow systems that achieved maximum occupancy once the number of beds to be staffed were properly determined. In 1974, one of Hancock's first IOE PhD students working in hospital systems engineering, James Martin, joined Griffin and Hancock in the School of Public Health as an assistant professor. In 1976, Griffin, Hancock, and Fred Munson published the first book on the topic, *Cost Control in Hospitals*, which was lauded for its use of case studies showing that a data-driven systems approach could effectively reduce the cost of many different types of hospital operations. With the advent of mini computers in the '70s, and with funding from a National Institutes of Health (NIH) grant, they developed a computer simulation meant to allow hospitals to maintain much higher occupancies, and at the same time provide for unexpected emergency arrivals. The first successful implementation of this simulation resulted in a six percent increase in patient occupancy with an increase in budget of only $10,000 annually in a large hospital. This was a time when NIH officials were promoting translational research. The project director of NIH encouraged Hancock to start a company to implement the Admission Scheduling and Control System (ASCS) in other hospitals. Hancock successfully implemented ASCS in 20 hospitals. In 1983, Hancock and Paul Walter published a widely acclaimed technical manual describing the process: *The ASCS: Inpatient Admissions Scheduling and Control System*. By this time Hancock had supervised seven IOE PhD students who contributed to a body of research on the use of various types of statistical and operations research methods to reduce hospital costs and to improve occupancy rates and patient care. One of their contributions was a computer-aided operating room scheduling system that enabled hospitals to increase utilizations by 20 percent and to start surgical procedures on time 95 percent of the time. This was implemented in six hospitals. Also, nurse staffing and scheduling algorithms were developed that allocated patients to nurses so that the probability that the nurse could do all of the work required was very high.

It was estimated that implementation of ASCS would result in a budget reduction of over 30 percent in the more than 4,000 secondary hospitals in the country, and at the same time result in much higher quality of care. By 1995, a total of 12 students received their PhDs from this effort, nine of whom were IOE students.

In 1984, Hancock shifted his primary work back to the College of Engineering when he became the associate dean for the manufacturing initiative, though he still supervised several PhD students in the health care area. His impact on manufacturing operations led to his becoming the first William Clay Ford Chair of Product Manufacturing in 1989. With the end of Hancock's leadership in healthcare systems during the '80s there was a period when the department continued to offer a course in hospital systems and encouraged many undergraduate students to engage in projects in the UM Health System, but the department did not lead any related research. For the next 15 years the hospital systems course and student projects were mostly taught and supervised by adjunct professor Richard Coffey, a PhD graduate of the IOE Department in 1974. Coffey was the director of industrial engineering in the UM Health Systems. Like many important and complex problem areas, however, interest in the topic of health care did not disappear within the department, as discussed in chapter 8.

[7]

The Center for Ergonomics

7.1. The Foundation Years for the Center for Ergonomics (1960s and 1970s)

The origins of ergonomics, (which means "the study of work" in Greek) within the Department of Industrial Engineering began with the formation in 1961 of the Engineering Human Performance Laboratory lead by Walton Hancock. As was described earlier, this laboratory was supported by the Methods-Time Measurement (MTM) Association, a consortium of companies that supported and conducted empirical research on worker productivity and the factors affecting it. Hancock believed both field studies and controlled laboratory studies were necessary to understand and quantify human behavior in various work environments. In pursuing the development of a human performance laboratory, he was joined by research engineer James Foulke in 1960. Over the next 10 years, with the support of the MTM Association and, later, the Chrysler and Ford corporations, Hancock, Foulke, and several graduate students provided a large amount of field and laboratory data describing worker movement times and the conditions that affected operator learning rates, error rates, and decision processes.

Physical fatigue and heat stress issues were studied in this laboratory during the mid-1960s by Don Chaffin, who at the time was an industrial engineering PhD student. After receiving his PhD in 1967, Chaffin joined the faculty in the Department of Physical Medicine and Rehabilitation at the University of Kansas Medical Center. He returned to the University in 1969 as an assistant professor in the IE Department. During this same period, Richard Pew, whose primary faculty appointment was in the Department of Psychology and who had expertise in human-perception and display design, was provided a joint appointment in IE. In 1971, James Miller, an expert on safety systems from Ohio State University joined the department as an assistant professor, further expanding the lab's scope to include safety engineering.

With the combined expertise of professors Hancock, Chaffin, Miller, and Pew, and with the successful operation of the Engineering Human Performance Laboratory, the first occupational safety engineering training grant was secured in 1972 with Chaffin as the project director. This grant from the National Institute for Occupational Safety and Health provided funding for six to eight graduate students annually who were studying in the industrial and operations engineering (IOE) subspecialty of ergonomics and safety engineering.

Also in1972, Chaffin became the director of the newly named Human Performance and Safety Engineering Laboratory, replacing Hancock, who had shifted his work to understanding and modeling hospital systems. A year later Gary Herrin joined the department from Ohio State University, adding expertise in statistical analysis and quality assurance methods. Herrin further provided technical support and leadership for a host of empirical studies of worker health and safety at various industrial sites around the country. In 1974, Gary Langolf, who earlier had received his PhD from the IOE Department, returned as an assistant professor after being on the faculty at Wayne State University for a couple years. His research concentrated on perceptual motor control related to precise movements. Langolf and Chaffin immediately started collaborations with Larry Fine, an occupational epidemiologist in the School of Public Health, and James Albers, a neurologist in the School of Medicine. Together they conducted fundamental studies that quantified how common neural toxic exposures in the workplace affected perception and movement control, with grant support from the National Institute for Occupational Safety and Health.

In 1977, Thomas Armstrong, who had received an interdisciplinary PhD in industrial engineering, occupational health, and physiology from the University of Michigan (UM), was appointed an assistant professor in the School of Public Health, with a joint appointment in the IOE Department. His research was primarily on understanding and preventing upper extremity cumulative trauma disorders in industry. Louis Boydstun joined the faculty as an assistant professor in 1977 from Purdue University. He specialized in human-display interface designs. Lastly, in 1980, Devinder Kochhar, an expert in visual display designs, joined the faculty from the University of Waterloo, replacing Boydstun, who left in 1983 to join the Jet Propulsion Laboratory. Thus, by the end of the '70s the IOE Department had five full-time ergonomics faculty members (Chaffin, Miller, Herrin, Langolf, and Boydstun, with part-time support from Armstrong, Fine, and Albers).

The '70s provided many opportunities for the faculty in the Human Performance and Safety Engineering Laboratory to find financial support for their research, as companies sought to improve the human aspects of work. During the '70s there was a growing concern among business management groups and unions to reduce the cost and suffering caused by occupational overexertion injuries. With the advent of the Occupational Health and Safety Administration in 1970, combined with better computerized record-keeping systems, worker injuries began to be publically reported, along with the costs of treatment and lost work time. This open reporting revealed that more than $100 billion annually was being spent in the United States on occupationally related lower back disorders, carpal tunnel syndrome, and many other types of musculoskeletal disorders; in some companies more than a third of the workers had injuries that resulted in medically prescribed work restrictions.

This era also saw people continue to travel into space, which brought NASA funding for Chaffin to conduct several studies to predict the human strength capability of astronauts when they attempted to perform manual tasks aboard the Skylab or on the lunar surface. Chaffin also teamed up with Richard Gerald Snyder, a professor of anthropology and director of the UM Transportation Safety Institute's Life Science Division, and Clyde Owings, a professor of pediatrics and bioengineering, to produce a strength prediction model of young children that was used to set safety standards for toys and other objects that could be hazardous when manually handled. In addition, Richard Snyder, Don Chaffin, and Rodney Schutz (a PhD student in IOE) also produced the first three-dimensional (3D) kinematic model of the human torso that was then used to predict maximum reach capabilities in vehicles.

Thus, by 1979, the Human Performance and Safety Engineering Laboratory was able to augment the annual National Institute of Occupational Safety and Health (NIOSH) Training Grant funding of $192,000 with contracts and grants of more than $600,000 from Western Electric, ALCOA, Kaiser, United Airlines, NASA, NITSA, NIH, Jewel Food Stores,

Owens-Corning Fiberglas, General Tire, Firestone, and Dayton Tire and Rubber Companies. This total annual funding would be the equivalent to about $3.2 million in 2014.

Based on the magnitude of the work being done, the five IOE faculty members associated with the Human Performance and Safety Laboratory sent a proposal to Dean David Ragone in 1979 to change the name of the laboratory to the Center for Ergonomics and to have the director report to the dean of the College of Engineering. This proposal indicated that the laboratory's combined external funding was providing partial support for 13 faculty members, six staff members, 33 graduate students, and six undergraduate students. The proposal also indicated that in the prior 10 years the lab had supported 130 students who had received MS degrees and 10 students who had received PhD degrees.

7.2. The Center for Ergonomics is Born

Given the success of the faculty members involved with the Human Performance and Safety Engineering Laboratory, the name Center for Ergonomics was confirmed by the regents of the University in November 1979. Gary Herrin served as the first director of the center from 1979 to 1981. After serving as chair of the IOE Department for four years, Chaffin became director of the Center for Ergonomic in 1982 and served in that capacity until 1998, when Thomas Armstrong, who had moved his primary faculty appointment from the School of Public Health a few years earlier, became the third director of the center.

Gary Herrin, PhD – 1979–1981

Don Chaffin, PhD – 1982–1998

Thomas Armstrong, PhD – 1998–2015

Education in ergonomics had expanded in the department during the late '70s, as indicated in the list of IOE courses shown earlier. Each faculty member developed specialized graduate courses in different aspects of ergonomics, and the department hosted formal seminars and workshops by experts from around the world in this growing intellectual discipline. Several introductory undergraduate courses were also developed for IE and other engineering students. That the quality of the educational program in ergonomics and safety engineering was recognized by the National Institute for Occupational Safety and Health is evident by the fact that they continued to renew their training grant for the next 35 years.

It was also about this time the faculty started offering a series of short courses for engineers and practicing safety and health care professionals. Les Galley was hired in 1982 as a projects manager and director of continuing education in the center. In this capacity, Galley helped organize five to nine short courses annually and helped coordinate the center's work with various companies. In 1985, Randy Rabourn took over Galley's position and, with the advent of the NIOSH Educational Resource Center Grant that had been established since 1982, became the director for the continuing education courses at the Center for Occupational Health and Safety Engineering, which at times numbered more than 20 courses annually. The ergonomics short courses were very popular, sometimes being attended by over 200 people, and were often offered three times each year throughout the '80s and early '90s.

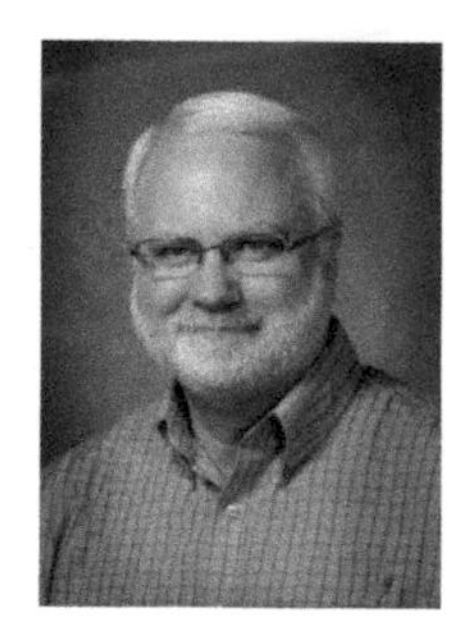

Charles Woolley (2011)

Randy Rabourn at retirement (2011)

As mentioned earlier, Jim Foulke joined the IE department in 1960 to develop advanced instrumentation as part of the Human Performance Laboratory. Later, in 1981, Charles Woolley joined Foulke in designing and building very advanced hardware and software systems to measure and analyze human performance attributes of all kinds, such as: metabolic energy rates, human muscle strength, eye movements, muscle contraction, and fatigue states via such means as electromyograms and motion capture. Foulke and Woolley also developed and taught courses describing contemporary human measurement systems.

Jim Foulke with early posture analysis system (1983).

Sheryl Ulin joined the center staff in 1985. Her expertise is in organizing and conducting field studies and using psychophysical methods to quantify discomfort levels. Ulin and Rabourn also established an ergonomics training and job analysis program for small to medium-size companies in Michigan, which has been continually funded by the state of Michigan since 1992. Finally, it is necessary to recognize Pat Terrell, who oversaw the production of a myriad of reports and grant proposals over 30 years with the center, as well as Kelley Cormier, the Center's Administrator for many years, Patricia Cottrell who worked closely with Randy Rabourn to organize many of the continuing education courses, and Richard Sullivan, who as Center Receptionist handled many different, daily requests.

Left to right: Richard Sullivan, Pat Terrell, Patricia Cottrell, and Kelley Cormier

Other faculty members who have played a major role in the center as research scientists are Laurence Fine and Alfred Franzblau, both occupational physicians and professors of environmental health sciences, and Robert Werner, professor and chief of physical medicine and rehabilitation at the VA Medical Center. As mentioned earlier, James Albers, a professor in the Department of Neurology, worked closely with IOE professor Gary Langolf to establish the behavioral effects of occupational exposures to various neurotoxins. These faculty members brought a much needed clinical component to the Center for Ergonomics. In addition, Paul Green, who had received a joint PhD in IOE and psychology from the University in 1979 and was working as an assistant research scientist at the Transportation Research Institute, started developing and teaching courses in ergonomics in the department. Many of Green's studies over the years have been done in collaboration with students and staff associated with the Center for Ergonomics. Currently, he is an adjunct associate professor of IOE and a research scientist in the Transportation Research Institute.

Paul Green, PhD

Robert Werner, MD

Alfred Franzblau, MD

Sheryl Ulin, PhD

James Albers, MD, PhD

Lawrence Fine, MD, DrPH

7.3 The Importance of External Shared Funding

7.3.1. The Ford Grant

By 1980, the Ford Motor Company, along with other domestic auto companies, was still struggling to recover from the ramifications of a smaller domestic car market and some well-publicized product quality and reliability issues. Ford executives believed a new management-labor paradigm was needed and decided to elevate the union's power through what they called employee involvement programs. As part of this trend, the Ford manufacturing executives sought the assistance of faculty members in the Center for Ergonomics. Ford agreed to provide the center with $400,000 annually for five years (the total grant amount would be the equivalent of more than $5 million in 2014). In return the faculty members would establish a series of ergonomics studies in several Ford plants to determine how changes in work conditions, tools, and processes could prevent worker injuries and improve product quality. Within a year eight PhD students were conducting research on ergonomics issues in the manufacture of Ford vehicles. The center's faculty also collaborated with Jeff Liker, an expert in organizational management who had joined the IOE faculty in 1982. This collaboration resulted in the production of training books and videos that the Ford management and the UAW then used to train engineers, safety officers, and union leaders on how to perform fundamental ergonomic studies in all their plants. Several years later GM and Chrysler used these training materials as the basis for the development of plant-level ergonomics programs that are still in use by many companies besides those in the automotive sector.

7.3.2. Expanding the NIOSH Training Grant

An important faculty addition occurred in 1984 when W. Monroe Keyserling joined the department as assistant professor of IOE. Keyserling had been serving on the faculty at Harvard University's School of Public Health after receiving his PhD from UM five years earlier. His research concentrated on improving the field methods used to evaluate occupational-related musculoskeletal stresses and safety hazards, postural fatigue, and slip-

and-fall hazards, which were of major concern in various manufacturing operations. Keyserling became the director of the NIOSH Occupational Safety Engineering Training Program shortly after arriving, and during the late '90s Directed the Center for Occupational Health and Safety Engineering, which in 1982 had become the UM umbrella organization for several NIOSH-sponsored occupational health and safety training grants in IOE, industrial hygiene, occupational medicine, and occupational health nursing. The grant for the Center for Occupational Health and Safety Engineering also provided support for a series of annual short courses for professionals in the field of occupational health and safety. The training grant had two effects on the center faculty. First, faculty from various departments had to meet often to organize and host common seminars, perform curriculum planning, and advise students with a variety of backgrounds and interests. Second, students in the various departments were required to take courses outside their own department. This created a successful multidisciplinary teaching and research environment that at times involved more than 25 faculty members from the four occupational health and safety disciplines.

7.4. Center Faculty in the 1990s—Growing the Perceptual and Cognitive Area

As described in chapter 6, some turnover in the IOE faculty affected the orientation of the research within the center. One such change occurred in 1986 when Jay Elkerton joined the faculty, replacing Devinder Kochhar in the area of perception and cognitive-related aspects of computer-generated displays. Kochhar had left to join the Bell Laboratory's human factors staff. Four years later, in 1991, Elkerton left to join the research staff at Hewlett-Packard. He was replaced by Yili Liu, who had acquired his PhD from the University of Illinois under the guidance of Chris Wickens, a highly regarded human factors and cognitive scientist. At the center, Liu established a successful research program on cognitive modeling and explored the importance of various cultural differences on the function and design of products.

Another personnel change occurred in 1989 when Gary Langolf was granted an early retirement. Shortly after, Bernard Martin from the National Institute for Transportation Safety Research in France joined the IOE faculty as an assistant professor. His research at the center focused on perceptual-motor control, especially in the presence of vibration, and in fundamental motion modeling. In 2004, Nadine Sarter joined the center from Ohio State University; she had demonstrated expertise in various types of auditory, visual, and tactile displays. Thus, by 2005, the center had six IOE faculty members, three in the physical ergonomics and biomechanics area (Armstrong, Chaffin, and Keyserling) and three in the perceptual and cognitive aspects of human-hardware systems (Liu, Martin, and Sarter), assisted by part-time faculty member Paul Green.

Don Chaffin

Tom Armstrong

W. Monroe Keyserling

Bernard Martin

Yili Liu

Nadine Sarter

7.5. Visiting Faculty and Research Scientists

The IOE Department faculty members in ergonomics have been privileged to be assisted over the years by a number of visiting faculty from various countries. Following are a few of the individuals who have added greatly to the intellectual and operational environment of the center.

Shrawan Kumar was a visiting professor from the University of Alberta in 1983–1984. Kumar is a highly regarded scientist and author on the topic of human physiology as it applies to ergonomics problems and overexertion injuries. Kumar is professor of physical therapy and neurosciences and a fellow in the Royal Society of Canada.

In 1984, **Hudson De Araujo Couto** was a visiting professor. Couto is the chair of the Department of Occupational Medicine at the Federal University of Minas Gerais, in Belo Horizonte, Brazil. His work in the center dealt with understanding and documenting the differences in physical work capacity between Brazilian and US workers involved in heavy manual labor.

In 1985–1986 the center hosted **Issachar Gilad** from the faculty of the Technion-Israel Institute of Technology in Haifa, Israel. Gilad is a well-known scientist in human motion analysis and biomechanics.

During the academic year 1988–1989 the center hosted **Owen Evans**, director of the Center for Ergonomics and Human Factors at La Trobe University in Melbourne, Australia. Evans is well known for his ergonomics publications on designing workplaces to be productive and safe for aging workers.

The center has also hosted a number of postdoctoral scholars, some of whom received their PhD degrees through research in the center. These included Don Bloswick (now professor of IE at the University of Utah), Arun Garg (now professor and chair of IE at the University of Wisconsin–Milwaukee), Mary Ann Holbein (now professor of kinesiology at Slippery Rock University), Maury Nussbaum (now chaired professor of industrial and systems engineering at Virginia Polytechnic Institute and State University), Rob Radwin (now chaired professor in IE and past chair of bioengineering at the University of Wisconsin–Madison), Mark Redfern (now professor and vice president of research at the University of Pittsburgh), Laura Punnett (now professor at University Massachusetts–Lowell), Zudong Zhang (now professor of mechanical engineering and orthopedics at the University of Pittsburgh), and Stephen Goldstein (professor of orthopedics and director of the Orthopedic Research Laboratory and later vice-provost for medical research at the UM and member of the National Academy of Engineering).

7.6. Impact of the Center for Ergonomics

The annual Ford grants, combined with funds from other companies and federal agencies, provided financial support that averaged more than $900,000 annually. These funds, and the NIOSH training grant funds provided the center faculty with the ability to attract and support about 15 to 20 PhD students annually throughout the ’80s and ’90s, as well as about 20 MS students. A total of 92 PhD degrees were granted to students who performed their research in the center between 1979 and 2005. Their research included such topics as an evaluation of various computerized ergonomic design methods, the effects of hand-tool vibration on tactility and discomfort, biomechanical models of the lower back when lifting, a computerized expert safety system, biomechanics of the hand and wrist during manual exertions, lower limb discomfort during prolonged standing, movement control in high-precision tasks, and multiple display scanning strategies. Appendix A.4 lists the names of those who earned IOE PhD degrees, including those obtained through research within the center.

It should also be noted that for more than 44 years NIOSH has continued to provide funding for the faculty and graduate students in the ergonomics program, making it not only the first engineering education program formally recognized by this federal agency in 1971 but also the longest continually funded program in ergonomics and safety. During this period the NIOSH program supported 234 students who were granted master’s degrees, and 66 students who earned PhD degrees in ergonomics and safety engineering. In addition, hundreds of other graduate and undergraduate students benefited from the rich diversity of ergonomics and safety courses provided by the faculty associated with the center over the years.

Many of the more than 20,000 alumni from both the academic degree programs facilitated by the center and from the many ergonomics short courses for practicing professionals have achieved notable leadership positions in industry, government, and academic institutions.

This is best exemplified in the US automobile industry, which is recognized throughout the world for its use of ergonomics to improve the health and safety of its workers and the quality of its products. Many graduates from the ergonomics programs have led this development, which has contributed to injury rates in automotive manufacturing plants dropping to less than one-third what they were in the early '90s. They have also provided much of the leadership now being sought by other industries that are striving to adopt effective ergonomics programs.

More than 1,300 papers, books, book chapters, and technical reports have been produced by the faculty and staff members associated with the Center for Ergonomics since its founding in 1979. *Occupational Biomechanics*, a textbook written by Don Chaffin, Gunnar Andersson, and Bernard Martin in 1984 is in its fourth edition and has been adopted by over 200 universities at various times, according the publisher, John Wiley and Sons.

Several software programs meant to assist practicing engineers and safety and health professionals have been produced by the faculty and staff of the center, such as the 3D Static Strength Prediction Program™ and Energy Expenditure Prediction Program™. The University has licensed these two programs to more than 3,000 individuals and companies around the world since the mid-1980s, and they remain as one of the first and most successful software technology transfer ventures that the University has undertaken.

7.7. The Human Motion Simulation Laboratory

Julian Faraway, PhD

The Human Motion Simulation (HUMOSIM) Laboratory was initiated in 1998 by Don Chaffin within the Center for Ergonomics to study and model how various people move and perform exertions in a large variety of manual tasks. The success of the HUMOSIM Laboratory was ensured by the intellectual leadership provided by Julian Faraway, from the Department of Statistics, and Matthew Reed, who was at the time an associate research scientist in the UM Transportation Research Institute. Faraway later accepted a chaired professorship at University of Bath in the United Kingdom, and Reed became a research professor and director of the Life Sciences Division within the UM Transportation Research Institute. Others from the center who guided early research in the HUMOSIM Laboratory were IOE professors Bernard Martin and Thomas Armstrong, and mechanical engineering professor Brent Gillespie. For the first seven years the lab's funding was provided by a consortium of companies and federal agencies. These organizations were involved in using human simulation algorithms to evaluate and improve their products and work environments. This consortium included Chrysler, Ford, GM, Toyota, Navistar, Lockheed Martin, the US Postal Service, the US Army TACOM (Tank-automotive and Armanments Command), and the UM Army Automotive Research Center. These groups provided more than $700,000 annually between 1999 and 2005.

Matthew Reed, PhD

Though several analytical methods were being used to simulate human motions, primarily for gaming and movie productions in the late '90s, the only criterion that seemed to apply to these was that the simulated motions had to appear to be realistic, though many obvious flaws were apparent when these early motion simulations were closely observed (e.g., feet that slid on the floor rather than being lifted while a character was walking or turning, eyes that didn't move with the hands, fingers that didn't deform when grasping).

Illustration depicting the use of the Jack digital design software from Siemens AG to evaluate a proposed new manual welding operation. Normal human motions from the HUMOSIM Laboratory were used to animate Jack and other human avatars imbedded in commercial CAD programs.

Considering that major engineering design decisions could rest on the simulations of people using and manufacturing products, a much more robust and accurate human motion technology was needed. In consultation with the sponsors, four criteria for human simulations were established to guide the activities within the HUMOSIM Laboratory:

1. Simulated motions must be based on real human motion data to have internal construct validity and empirical validity.
2. Models of motions should be able to represent motions not in an existing database; that is, they should have extrapolation capability.
3. Models should be computationally fast and portable for real-time simulations.
4. Models should automatically adapt and use new motion data to become more robust in predicting novel motion situations of interest to a designer.

Meeting these criteria requited the running of more than 200,000 laboratory studies of different people reaching, grasping, moving objects, walking, and carrying objects. Sophisticated 3D motion-capture systems were used to document these motions, and the resulting kinematic data were placed in a standardized and well-documented database created and maintained by Charles Woolley. The sponsoring organizations were granted free access to the database, as were other organizations that wished to model the data. By 2000, the HUMOSIM Laboratory was supporting 12 PhD students and six MS and BS students annually. Models were created that used a variety of functional regression, kinematic optimization, and biomechanical methods to meet the four criteria.

Between 1998 and 2005, 96 papers and technical reports were published on the lab's work. Most importantly, under the direction of Reed a general-purpose kinematic framework was created that allowed the modeling results to be used to predict a wide variety of common motions. By referencing this framework, users of a commercial computer-aided design (CAD) system that included a human avatar could access a diverse set of data-driven motions in their CAD simulations, thus forgoing the need to perform additional time-consuming and expensive human motion laboratory studies. For more information about the HUMOSIM Laboratory, see its website: www.HUMOSIM.org.

7.8. The Rehabilitation Engineering Research Center

Design of equipment and tasks to accommodate persons with the broadest range of physical and mental capacities is an important application of ergonomics. The Center for Ergonomics has conducted two major rehabilitation projects. First, the Michigan Bureau of Rehabilitation Services and General Motors supported work concerned with development of workplace accommodations from 1978 to 1983. A field team from the center visited a number of Michigan industries with a mobile laboratory to demonstrate tools and procedures for analyzing jobs and identifying possible employment barriers. In addition, the project offered a number of short courses to train rehabilitation counselors from the state of Michigan and safety, health, and rehabilitation coordinators from Michigan employers. This project brought together the faculty and a number of students affiliated the Center for Ergonomics in one way or another at that time. Arthur Longmate, a PhD candidate in IOE coordinated the field visits and data collection with the mobile laboratory. Don Chaffin, Gary Herrin, and Monroe Keyserling worked on developing strength testing protocols and determining applications for early whole-body biomechanical and metabolic prediction models. Louis Boydston worked on developing models for describing 3D reach capacities for seated workers, and Tom Armstrong coordinated the overall project.

From 1997 to 2002, the Center for Ergonomics was home to a Rehabilitation Engineering Research Center (RERC) supported by the National Institute for Rehabilitation Research. The RERC utilized a web-based model to focus research on developing tools to determine job demands and worker capacities, identify gaps in how job-worker matching was being done, and develop interventions to better match a person's capabilities with job demands. The RERC emphasized primary and secondary rehabilitation of persons with chronic musculoskeletal disorders. It brought together investigators and students from the Center for

Ergonomics (Armstrong, Chaffin, Martin, Keyserling, Koester, Ulin, Rabourn, Foulke, and Woolley), the School of Public Health (Franzblau), and the Department of Physical Medicine and Rehabilitation (Haig, Werner, and Levine) who collaborated on a number of research initiatives and demonstrated their applications. The RERC was codirected by Armstrong, Haig, Levine, and Werner. The RERC also helped to support an active outreach and continuing education program at a national level (Coordinated by Rabourn and Ulin). The Center for Ergonomics facilitated the research, which involved faculty members and students from the IOE Department, School of Public Health, and Department of Physical Medicine and Rehabilitation. A number of field and laboratory studies were completed over the period of the grant that provided new knowledge and tools for preventing work disability due to musculoskeletal disorders. A website (see the image of the original home page) was developed to provide RERC resources, including an extensive job database available to others working in the field.

Contact Us
Project Participation
Project Personnel
Research Projects
IRB
Courses & Meetings
Meeting Schedule
RERC Advisors
Related Links
Work Design Examples
Examples
Home

University of Michigan
Rehabilitation Engineering Research Center

Ergonomic Solutions for Employment

Objectives

- Develop a Model System for applying ergonomic technologies to accommodate disabled and elderly workers. Include procedures for:
  - assessing workers
  - identifying accommodation needs
  - selecting ergonomic interventions
  - Integrate findings from Intervention and Biomechanics Projects into Model System
  - Implement a database to evaluate the system and make it available to others

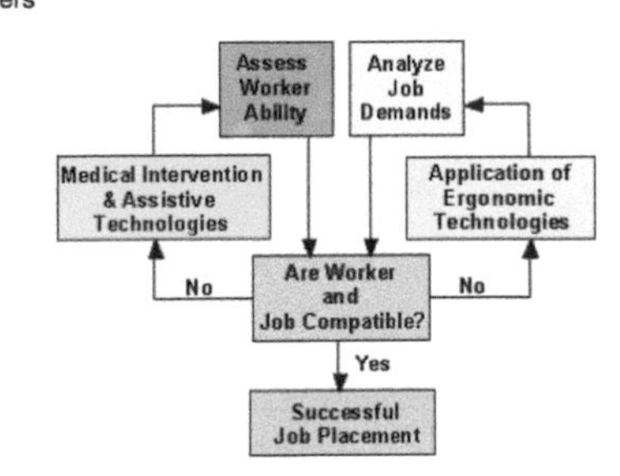

- Analyzing jobs
- Identifying and rating ergonomic barriers to employment
- Case Examples

Participating University of Michigan Departments:

- Industrial and Operations Engineering
- Physical Medicine and Rehabilitation
- Environmental and Industrial Health

7.9. Some Final Comments about the Center for Ergonomics

It is worth noting that all of the faculty members associated with the Center over the years have maintained a high level of scholarship. More than 1,300 ergonomics reports, journal articles, papers in proceedings, book chapters, and books have been published between 1979 and 2005. Such an archive of ergonomics knowledge, most of which is available through the center's website, continues to be recognized by other organizations. The faculty associated with the center have also received many distinguished honors and awards for their work from a large variety of scientific and professional organizations throughout the world. Most importantly, many of the graduate students who have passed through the center have gone on to distinguished careers in academia and industry.

In summary, in many different ways the Center for Ergonomics has provided the knowledge, tools, and trained professionals needed to improve the well-being and safety of countless numbers of workers and consumers throughout the world. It has been an exciting time for the center, and this continues to be the case as the 21st century appears to have brought a growing public interest in convenience and safety in all types of products, services, and production facilities. Much has been accomplished within the center, but much remains to be done. The IOE Department has provided an excellent home for the center over the years, and though many faculty members from other departments have participated in the research and educational programs supported by the center, the organizational and administrative support provided by the IOE Department and the College of Engineering has allowed the center to thrive. In essence, over 44 years, the many outstanding accomplishments of the people associated with the center are proof that the University of Michigan's policy of actively encouraging groups of faculty members from various departments, colleges, and schools to work together on large and important societal issues has been extremely valuable to the nation.

[8]

Epilogue

Though this book is about the events and people during the first 50 years of the industrial and operations engineering (IOE) Department (1955 to 2005), it is worth noting that it is being written in 2015. Some have asked why I have not included this recent 10-year period in the book. The simple answer is that I believe it is too early to know what these past 10 years will offer that is historically important. Having said that, I do think that it is worth noting some important national trends that appear to have affected the department over the past decade and may present significant challenges and opportunities in the future.

8.1. IOE and Health Care Systems Engineering

As discussed in chapter 6, in the '60s, '70s, and early '80s, the IOE Department was a pioneer in the application of industrial engineering expertise to improve the management of hospitals and health care systems. Though the research base for such activity in the department diminished in the late '80s and early '90s, at the same time there was increasing interest and financial support at the national level for developing engineering solutions to biological and medical problems. This resulted, for instance, in the creation of many new bioengineering departments across the country. At the University of Michigan (UM) the Bioengineering Interdisciplinary Graduate Program that had been serving as the home to more than 50 master's and PhD students annually since the early 1960s, was transformed in 1996 to support a BS degree. By doing so this long-standing and acclaimed graduate program became the Biomedical Engineering Department in the College of Engineering. Several faculty members in IOE held joint faculty positions in this new department, namely Don Chaffin, Tom Armstrong, and Bernard Martin. The existence of this new department, the first new department in the College in 20 years, facilitated a larger conversation regarding the use of engineering technologies and methods to improve health care and medical practice in general as well as how faculty members outside of the Biomedical Engineering Program per se could be engaged in such activities.

Beginning around 2005, with active encouragement from College of Engineering dean David Munson and Health Systems dean James Wolliscroft, several faculty members in IOE formed partnerships with different medical school faculty members to study a large variety of medical and health care problems. One example is a series of studies conducted by Tom Armstrong's group within the Center for Ergonomics to carefully document the micromotions that are required of skilled surgeons during various surgical procedures for the purpose of improving the training of future surgeons and the design of the instruments they use.

Nadine Sarter's research group in the Center for Ergonomics provides another example. She and her students have been studying how the performance of anesthesiologists during surgical procedures can be improved using a variety of audio, visual, and tactile sensory input devices. Also, Marina Epelman has been working with the Radiology Department to determine the most effective method for scheduling follow-up examinations to detect and treat cancer patients.

A recently hired associate professor, Brian Denton has been working with the Department of Urology to improve the early detection and treatment of prostate cancer. Mark Van Oyen and Mariel Lavieri have been developing stochastic models for monitoring the progression and control of risk factors for glaucoma, as well as to predict mortality risk based on admissions and use of intensive care units and other forms of intermediate care units.

Richard Hughes, who earned his PhD in the IOE Department in 1991 and then served as an ergonomist for the state of Washington and a research scientist at the National Institute of Occupational Safety and Health and the Mayo Medical Centers, returned to the UM Department of Orthopedics in 1998. He is associate professor and codirector of the Michigan Arthroplasty Registry Collaborative Quality Initiative and directs the Optimization and Computation Laboratory in Orthopedic Surgery while holding secondary faculty appointments in biomedical engineering and IOE. Finally, Mark Daskin, the IOE Department chair is working on several health care projects, including one on nurse staffing and another comparing the goals and expectations of kidney transplant recipients to those of the patients' physicians, nurses, social workers, and other caregivers.

These individuals and others in IOE Department have managed to develop important partnerships with colleagues in the medical school. To further develop these established investigations in the medical arena, the department hired a distinguished professor of practice, James Bagian, who was a scientist astronaut with NASA, and had developed and led the VA Patient Safety Research Center for more than 12 years. He recently teamed with IOE professor Amy Cohn to form the Center for Healthcare Engineering and Patient Safety. The College has provided support staff to nurture this new endeavor, and a few postdoctoral students as well as a large group of undergraduate and first-year graduate students from IOE and other university departments have become involved. It is too early to determine if this new organization can provide the type of infrastructure and financial support needed to have a significant and lasting impact on medical practice, but clearly, a large number of IOE faculty members are now actively engaged in important health care and medically related work. If these faculty members can work together and are able to secure further financial support for this new center, as was done in establishing the Center for Ergonomics and the Information Systems Design and Optimization Systems program, then this activity could become a major national resource and have a significant and lasting impact.

8.2. IOE and Robotics and Autonomous Vehicles

Another national trend in the past 10 years that is relevant to the University and the IOE Department is the growing capability and use of sophisticated technologies meant to augment the performance and improve the safety of vehicles. For instance, technologies now exist and are being deployed that will detect and automatically stop a vehicle that is approaching another vehicle or object at too fast a velocity. Lane-following technologies are now able to keep a vehicle from driving off a road, and cars can now parallel park themselves. Trucks on open highways are beginning to travel without the driver's having to provide steering and braking functions. Navigation systems on board vehicles are providing instantaneous instructions about traffic conditions that could affect the selection of routes.

These are all technologies that exist and will facilitate the further development of completely autonomous vehicles in the near future.

The University of Michigan Transportation Research Institute (UMTRI) has been highly involved in studying how these various technologies affect the safety and performance of drivers. Many of the present autonomous vehicle developments rely on advanced robotics technologies. In essence, the autonomous vehicles of the future can be considered to be much like a robot in that the driver simply gives the vehicle a task to perform, for example, drive quickly and safely from point A to point B. To perform this task the vehicle then chooses the best route, guides the vehicle along the route, monitors the vehicle and environment, and communicates with other nearby vehicles. Much of this is accomplished using algorithms that are common to robotic developments. At UM this realization has resulted in a new multidisciplinary master's degree in robotics and autonomous vehicles within the College of Engineering.

One of the challenges in developing new autonomous vehicles is to ensure that the driver is able to maintain situational awareness while participating in other activities (e.g., talking to passengers, reading, conducting phone conversations, listening to engaging music) while the vehicle is in its autonomous driving mode. Some IOE faculty members are directing research related to this driver-vehicle control problem. For instance, Nadine Sarter, an expert on the use of multisensory displays, is studying how to improve the vehicle's communication with the driver, for example, warning the driver of potential hazards while not being overly disruptive and confusing. Paul Green and his group at UMTRI have been studying how drivers can be distracted by various navigation and entertainment systems to the point that they are not aware of immediate hazardous situations. Yili Liu has developed cognitive models of drivers that predict conditions that cause a mental overload, which can adversely affect decision making. How might this apply to future drivers? Consider a situation in which the driver of an autonomous vehicle is performing various tasks other than driving the vehicle but is suddenly required to take control of the vehicle. This sudden change in attention and cognition can easily cause a mental overload and degrade the performance capability of the driver. In other words, the control of the vehicle, whether by a human operator or the on-board computer, must be carefully managed under a variety of common and not-so-common driving scenarios.

Another example where the IOE faculty members are involved in future vehicle designs is in the work of a recently hired assistant professor, Clive D'Souza, who has an interest in how various types of autonomous vehicle technologies will assist drivers and passengers who have physical impairments. Also, Matt Reed, an IOE research professor and director of the Life Science Division of UMTRI has been developing digital human models of drivers to improve the placement of vehicle controls, seats, restraint systems, and airbags.

Clearly, there are many opportunities for collaboration between those on campus who are involved in the research needed to improve the transition to more autonomous vehicles. The long established working relationships between the University and automobile companies should provide the foundation for support of expanded activity in this problem area. It certainly is an exciting and important area of research and practice, and one in which the

IOE Department could become even more active, especially with its renowned expertise in ergonomics and large-scale, dynamic systems optimization.

8.3. IOE and Big Data

Huge collections of human-generated data are already being dwarfed by machine-generated data sets, many of which are orders of magnitude larger. A recent information industry report estimates that by 2020 more than 5 petabytes (500,000 gigabytes) will be stored worldwide for every man, woman, and child on the planet

This situation presents many opportunities and challenges. The opportunities are related to the goal of improving the efficiency and operations of economies, businesses, and government agencies by exploiting information contained in these massive data sets. In essence, it is broadly believed that improving the use of these massive databases will enable better decision making in a wide range of situations, for example, financial, medical, energy management, and security. One of the major challenges, however, appears to be that the expertise and methods to realize this goal are currently lacking.

What types of expertise are needed? Some believe the expertise is of the type that will make the data more understandable by human decision makers through innovative visualization systems. Resolving this problem will require a deep understanding of a person's perceptual and cognitive capabilities and the decision-making requirements posed by a task involving large and diverse data before developing a data presentation format. For example, three-dimensional (3D) visualization of unfamiliar city streets can be more helpful in finding destinations than text or audio instructions telling a driver where to turn.

Sometimes existing large databases are used as input to fully automated systems, combining past data analysis models with real-time control systems, as in the case of autonomous vehicle systems. A common example of such blending of past and present data is provided by systems controlling city traffic, where real-time adjustments to the timing of traffic light changes meant to reduce congestion is based on historical traffic data models and immediate traffic conditions. However, there are numerous examples of possible risks when automated decision systems lack adequate controls and safeguards. In this context, there are many publically reported problems associated with using forecasting models derived from past data, such as the financial disaster of 2008. It has been largely accepted that a major cause of the resulting financial recession was related to professional financial analysts relying on sophisticated market forecasting models that were not appropriate for the conditions at that time.

There have also been publically reported situations when essential data were not provided in a timely manner, such as a recent situation in a Dallas hospital when there was a delay of several days in correctly diagnosing and treating a patient with Ebola virus because the attending physician did not have access to data showing that the man had recently returned from West Africa. The emergency department nurse had entered the information in the man's record, but that particular part of the record was not available to the attending physician. The result was that many people needed to be quarantined and some required extensive medical treatment.

Technology for measuring, collecting, storing, securing, and presenting massive amounts of data about a broad spectrum of activities now exists, but if we are to effectively make use of the information contained in the data, we need advances in several areas:

- Combining and resolving conflicts present in multimodal data: Multimodal refers to data coming from different sources and different formats. If we wish to know whether a car with New York license plate ACE 8740 drove down Fifth Avenue in New York between 5:00 and 6:00 pm on December 1, we must combine seven characters of text with millions of pixels of video data extracted from a video database based on time and location.
- Presenting data in a usable form: A high-precision 3D image of a tumor can be much more helpful to a surgeon than an x-ray image, but the x-ray image is much more helpful than just a general description of the tumor size and location. A researcher investigating tumor growth, however, may require just the size and location.
- Maintaining the integrity of data: Data are inherently fragile and can be damaged accidentally or malevolently. Massive cases of identity theft are becoming increasingly common. How can algorithms be developed that more quickly and accurately identify and notify appropriate personnel of deviant data uses?
- Analyzing data to find the answer to a question: An ever-growing population of algorithms is being proposed for analyzing (e.g., mining) data. All have strengths, weaknesses, and limitations. Many work on some types of data but not on others. How do we decide which algorithms should be used for a specific situation? Can new and better methods be developed to avoid a false result from a chance correlation of large amounts of data being analyzed?

It is believed that we are just beginning to frame some of the issues in a way that can lead to better data-driven decision making in the future. But to move forward will require a multi-disciplinary, systems of systems approach. Such a broad approach was advocated by Yong Shi in a recent series of articles in the National Academy of Engineering's *Bridge* on big data, in which he stated that the necessary work to resolve the complexity of the problems "will involve interdisciplinary effort from mathematics, sociology, economics, computational science and management science" (*The Bridge*, Winter 2014).

When dealing with complex systems, improving decision making will require a much better understanding of the interactions that occur in and among at least three different domains illustrated in the diagram.

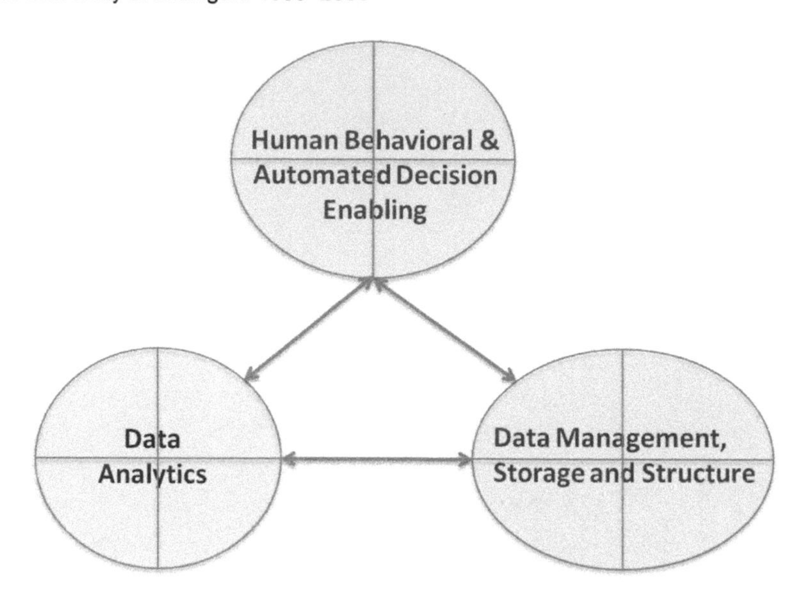

Different types of expertise required to improve data based decisions

At the University of Michigan a new multidisciplinary Data Science Initiative has been developed. This program envisions the establishment of a Michigan Institute for Data Science and a Data Science Services and Infrastructures. The former would be dedicated to developing expertise in handling crosscutting challenges, such as the following, to better serve a broad range of data science activities:

- Data integrity and provenance
- Integration of heterogeneous data
- Temporal and multiscale methods
- Data quality and uncertainty
- Data analysis and inference
- Data privacy and security
- Visualization of complex data.

The proposed Data Science Services and Infrastructure group will coordinate and publicize related courses and instructional projects and personnel, as well as work with other institutions. The expectation of the faculty members developing this organization is that eventually almost all disciplines in the University will be affected by the massive amounts of data that will be available in the very near future. Thus, there is a need for a formal organization to ensure that researchers and students have the intellectual support necessary to perform in this rapidly developing environment. The very breadth of the UM intellectual enterprise, with its 100 departments and 250 degree programs, provides the basis for believing that the University can achieve a leadership position in this arena if it marshals the right type of resources. Already federal agencies, such as the National Science Foundation, National Institutes of Health, Defense Advanced Research Projects Agency, and others, as well as some large foundations, are providing hundreds of millions of dollars to support research and instructional development related to this area.

One of the major spokespersons in this initiative is IOE and statistics professor Vijay Nair. Also, IOE research professor Marlin Thomas was hired recently to assist in developing a larger initiative for the department in this important arena. Though too early to predict, it is clear that this problem area could present many new research opportunities for IOE faculty members and students. While other disciplines tend to focus on collecting data; cleaning, accessing, and visualizing large databases; and predicting future trends (predictive analytics), there is a clear role for faculty and students in the IOE Department to carry these analyses to the next logical step—prescriptive analytics. This entails using big data to make better decisions. Both fundamental and applied research are needed in this area. For example, while advances in algorithm design and enhanced computing power now enable us to solve many hard problems to optimality in practice, massive amounts of data make some easy or polynomial solvable problems impractical. New algorithms will be needed to handle problems of this sort. Also, with massive amounts of data, data mining techniques run the risk of finding statistically significant results that in reality have no underlying basis in science or engineering. For example, if a data mining effort tests 100,000 hypotheses, at a .05 level of significance, we would expect 5,000 significant results, even when none really exist. Developing new statistical methods for such situations is another area in which the IOE Department can contribute.

8.4. Production Systems Engineering and Management

One of strengths of the IOE Department over its first 50 years has been in its ability to contribute to the fundamental knowledge and education required for improving a variety of production systems and their management programs. Most recently, as described in chapter 6, since 1991 the Tauber Manufacturing Institute (TMI) has become a national leader in providing master's-level education and field projects in manufacturing for IOE and School of Business students. This program has continued to grow as students realize the importance of having combined engineering and business backgrounds and degrees. IOE professors Larry Seiford and Yavuz Bozer have served as codirectors of this important graduate program. Bozer has also continued his work on the design of manufacturing and warehousing systems to improve parts flow and flexibility in low-cost distribution systems. Also, Judy Jin, who joined the department in 2005, continues her work in designing products to be manufactured in a manner that ensures high quality and reliability. Xiuli Chao, who joined the IOE faculty in 2003 from North Carolina State University, has continued his work in stochastic modeling and queuing systems related to production scheduling, inventory control, and supply chain management. Jack Hu, a recently elected member of the National Academy of Engineering who has led two major research centers in mechanical engineering that focused on improving the processes required to produce high-quality, assembled metal products, now holds an appointment in IOE. Hu is also serving as the university's vice president for research. His work links various IOE production planning and control methodologies to different types of manufacturing processes.

On the management side, Jeff Liker continues to study and compare how various international companies structure their production processes and procedures, as well as train their workers on the newest methods needed to produce high-quality products and services.

Several IOE faculty members in the operations research and ergonomics areas, whose research has been described elsewhere, have continued to work with manufacturing firms to improve their operations. It is clear that the manufacturing sector that provided so much of the foundation for the department is not being ignored and should continue to be the source of important problems and funding. This is especially true today as President Obama's administration pours millions of dollars into the establishment of manufacturing innovation institutes, one of which is located in southeast Michigan. Its research director is Alan Taub, a professor of practice in the College of Engineering. This 2014 funded institute is organized to bring together the best researchers and engineers from various universities, companies, and government laboratories to ensure that the United States will become preeminent in the field of lightweight and modern metals fabrication. Both Congress and Obama's administration are planning on funding up to 45 such institutes to tackle important problems in modern manufacturing systems. The IOE Department would appear to be well positioned to take advantage of these developing manufacturing innovation institutes.

8.5. Faculty Changes and Enrollments in the Past Decade

Some significant changes occurred in the faculty over the past decade. Some of the major early leaders in the department have retired in the past 10 years, namely Steve Pollock, Don Chaffin, Bob Smith, and Katta Murty. Also, some senior faculty members accepted positions elsewhere, such as Chelsea White, Chien-Fu Jeff Wu, Jan Shi, and H. Edwin Romeijn, who joined other former IOE faculty members at Georgia Tech and Georgia State University. We also lost John Birge to a deanship at Northwestern University and Jim Bean, who is now the provost at Northeastern University. In addition, Gary Herrin unexpectedly passed away in 2014. These senior faculty members had provided a great deal of visibility and influence within the department through their scholarship and professional service in their respective areas of interest. Also, with the exception of Murty, who made his contributions through his many books related to optimization methods and uses, it is interesting to note that these senior faculty members led several highly collaborative groups. The contributions of these various groups was lauded in all the departmental reviews over the years, not to mention that their collaborative efforts provided the source of a great deal of external funding and graduate student support.

On the positive side, more recent senior faculty members have joined the department.

As mentioned in chapter 5, Larry Seiford joined the IOE Department in 2000 as professor and chair. He was joined soon after by several faculty members who have been promoted and given tenure, namely, Amy Cohn, Nadine Sarter, Judy Jin, Marina Epelman, and Mark Van Oyen. Shortly after 2005. tenured professors Xiuli Chao, Edwin Romeijn, and Jon Lee joined the department. Seiford's successor as department chair, Mark Daskin, was hired in 2010. Daskin had served as the chair of the Department of Industrial and Management Science at Northwestern University. Others who were hired after 2010 and hold tenured positions in the IOE Department are Brian Denton, Pascal Van Hentenryck, and Seth Guikema. Other assistant professors were hired during this 10-year period, including Cong Shi, Siquian Shen, Viswanath Nagarajan, Clive D'Souza, Mariel Lavieri, Eunshin Byon, Henry Lam, Matt Plumlee, and Ruiwei Jiang. (See Appendix A.2 for a list of all faculty members and their areas of expertise). It should also be noted that two very distinguished, nontenured professors of engineering practice joined the department on a part-time basis: James Bagian (who was mentioned earlier) and Larry Burns. Burns, who had been vice president for research and development and strategic planning at GM Corporation, has become a major advocate for further development of better clean energy systems, particularly in the transportation sector.

During the previous 10 years there continued to be a growing need for lecturers and adjunct faculty members to meet the expanding teaching requirements, primarily, in the undergraduate program. See the chart for statistics on the number of tenured and tenure-track faculty (top row) and the number of lectures and adjunct faculty members (bottom row) in the fall semester each year.

Headcount	**FA06**	**FA07**	**FA08**	**FA09**	**FA10**	**FA11**	**FA12**	**FA13**	**FA14**
Total Faculty	24	23	23	21	22	24	23	24	25
Lecturers	2	2	3	4	4	8	7	5	6

It also is interesting to note that the average percent of an IOE faculty member's salary that is provided by the University's general fund has remained between 83 percent and 86 percent for the past decade. This is much higher than the average 73 percent in the 1980s, a period when the total research funding within the department was much larger and the undergraduate student body was much smaller.

The continuing growth in the department's undergraduate enrollment is shown by the graph that displays graduation rates from 2004 to 2013. This graph also displays a strong continuing interest by students in the department's master's degree. One potential concern is the relatively small number of PhD graduates over the years, but this trend will certainly improve as the department will be matriculating 26 new PhD students in 2015—the highest number of new PhD students admitted to the department in one year.

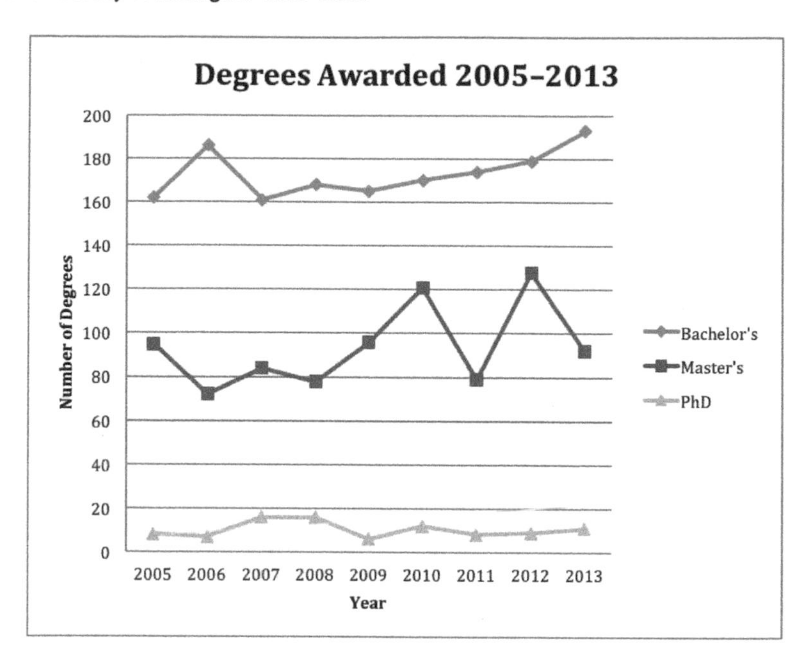

Several major factors appear to continue to contribute to the quantity and quality of the IOE PhD program that are worth acknowledging. First, the IOE Department's national ranking has remained exceptionally high during this past decade, typically number one, two, or three out of about 100 national programs, depending on the methodology and category used. This enables the department to attract excellent students at all levels. But PhD student education is expensive, often exceeding $100,000 annually per student. As discussed earlier, this requires faculty members to be very aggressive in seeking funds for themselves and their students. Another issue that affects the PhD program is the need for all the faculty members to be on the cutting edge of new and important problem areas. Such problems are often identified and studied within interdisciplinary groups, which requires many different types of collaborations to be formalized and publicized so that prospective students can easily identify their existence when considering their choice of where to apply for the PhD. The very active and formal faculty collaborations described in the previous chapters, which often provided the intellectual and financial support for much of the department's PhD program, are proof of the importance of such entities, but what should also be clear is that these types of organized groups should not be considered permanent. In fact, one measure of success for such collaborative efforts is how well they are mimicked and then improved upon by other organizations. As an example of this phenomenon, consider the Information Systems Design and Optimization Systems (ISDOS) organization in the '60s and '70s. As a pioneering effort in software development, it was influential in establishing several computer science programs at other schools. Those schools took the ISDOS model and improved it, while the commercial part of the ISDOS venture did not support the changes necessary to keep the UM research organization viable. The Center for Ergonomics provides a more current case.

When founded in 1979 it was the first such large research center in the United States. The public and institutional concern in the '70s over the growing cost and number of workers suffering from a variety of occupational musculoskeletal injuries propelled the center into being the leading research center focused on the use of engineering concepts to prevent such costly and disruptive disorders. This nationally important problem, which was recognized by many companies and government agencies, provided the primary sustenance for the center for the next 30 years. As discussed in chapter 7, however, though such physical disorders still exist in the workplace, the concerns over designing more complex systems and products that are more convenient, efficient, and safe to use by a variety of people has become an even more predominant concern in the field of ergonomics. Fortunately, as also described in chapter 7, the center is adapting to these changes with new and active collaborative relationships in health care and autonomous vehicle research.

On the other hand the Financial Engineering Program, which brought together faculty members from several different departments, has been largely usurped by similar programs at other universities that had strong existing ties with large financial institutions. This has not been the case for the TMI graduate education program with the Ross School of Business. This institute was formed in 1991 with major financial support over the years from Joel Tauber. As described earlier, it continues to provide excellent engineering master's degrees. In fact, it is estimated that over the past 24 years, 587 engineering students have received their master's degrees through the TMI. However, the institute has only supported a few PhD students. So the question must be posed; after 24 years is it enough to rest its future success solely on its abilities to provide superior education to students at the master's level? Or, putting it another way, how long can this or any other academic program maintain its high ranking within this research university without fostering and supporting an equally strong research program?

8.6. What's Next for IOE?

If one believes discussions in chapters 6 and 7 that collaborative efforts provide the vibrant and productive research and educational environments needed to achieve a real and lasting impact in science and engineering, then where are such new collaborations going to best develop in the department? The traditional development of methods to improve production and distribution systems within the department will surely continue. But my guess is that the health care systems engineering and robotics and autonomous vehicle areas, which are already providing exciting new opportunities for many IOE faculty members, will continue to greatly influence the department's future. In addition, the recent development of a coherent, multidisciplinary approach to data science and the role of IOE faculty members in this important but fledgling effort could provide great opportunities for additional IOE faculty members. Similarly, the organization on campus of several formal groups to coordinate research and educational programs dealing with new clean-energy systems is another recent and important phenomenon in which IOE faculty members may become more involved, as shown by IOE assistant professor Eunshin Byon, who is now conducting research in this area.

So to finish this epilogue, as Pat Riley, head coach and team president of the champion Miami Heat basketball team, once said, "Excellence is the result of always striving to do better." It is my hope that by providing this history of the IOE Department others will see more clearly what was necessary to achieve past excellence and what will be necessary to develop new areas of excellence in the future. In writing this book, and from being part of the department for more than 50 years, one thing has become clear to me. Whatever is done in the future, respectful collaboration will be necessary, as the problems we face are large and complex. In my opinion, no one person will be able to produce a major impact on these problems without the assistance and active support of colleagues, and achieving this goal will often involve working with those who have different types of expertise. If nothing else, I hope this book serves to inspire others to reach out and develop productive and long-lasting collaborations across this wonderful and eclectic campus.

Bibliography

Related Books on Early Industrial Engineering and UM History

Harold T. Amrine. Industrial Engineering at Purdue University—The Roots and First Thirty Years. West Lafayette, IN: Purdue University Press, 1984.

Anne Duderstadt. The University of Michigan College of Engineering—A Photographic Celebration of 150 years. Ann Arbor: The Millennium Project, University of Michigan, 2003.

James J. Duderstadt. On the Move—A Personal History of the University of Michigan's College of Engineering in Modern Times, Ann Arbor: The Millennium Project, The University of Michigan, 2003.

Frank B. Gilbreth and Lillian M. Gilbreth. Fatigue Study—The Elimination of Humanity's Unnecessary Waste. New York: Sturgis and Walton, 1916.

Lillian M. Gilbreth. As I Remember—An Autobiography. Norcross, GA Engineering and Management Press—Institute of Industrial Engineers. 1998.

Ernest J. McCormick. Human Factors. New York: McGraw-Hill Book Company, 1957.

Phillip M. Morse and George E. Kimball. Methods of Operations Research. first published by: MIT Press and John Wiley and Sons, New York, 1951, and reissued by Dover Publications, Inc. Mineola, NY, 2003.

W. Allen Spivey and Robert M. Thrall. Linear Optimization. New York: Rinehart and Winston, 1970.

Frederick W. Taylor. The Principles of Scientific Management. New York: Harper Bros., 1911. (available online from several sources)

Other References (Available for Inspection at Industrial and Operations Engineering Department)

Departmental Reviews: performed approximately every five years since 1968.

Department newsletters: published annually or biannually since 1961.

Faculty Meeting Minutes.

Related Contemporary Websites

Industrial and Operations Engineering Department, University of Michigan (http://ioe.engin.umich.edu/)

The Center for Ergonomics, University of Michigan (http://C4E.engin.umich.edu/)

Appendix

A.1. Wilbert Steffy and Outstanding Alumni Merit Award Winners

The Wilbert Steffy Distinguished Lecture

The Wilbert Steffy Lectureship was established in 2003 to honor one of the Industrial and Operations Engineering Department's early distinguished faculty members, Wilbert Steffy, who retired on May 31, 1976, after 29 years of service on the faculty of the College of Engineering. Steffy developed instruction in the area of cost analysis, engineering economy, and industrial purchasing policies. His association with the Industrial Development Division of the Institute of Science and Technology resulted in 10 widely distributed monographs on the application of industrial engineering and cost analysis to aid small businesses. He supervised many students and small businesses in his work, providing each of them with firsthand field experience in the practice of industrial engineering.

2004

Seth Bonder (shown on left) with presenter Larry Seiford

Seth Bonder, Improving Operations Research Support for Health Care Delivery Systems: Guidelines from Military Operations Research Experience

Seth Bonder was a leader in applying operations research to national defense planning and policy issues and subsequently to health care delivery reengineering and disease management practices. He served as president of the Operations Research Society of America and the Military Operations Research Society and was vice president of the International Federation of Operational Research Societies. His awards included the Patriotic Service Award from the secretary of the Army, the George E. Kimball Medal, and the Institute for Operations Research and the Management Sciences President's Award. The Seth Bonder Foundation continues to support students and research in operations research today.

2005

Thomas L. Magnanti

Thomas L. Magnanti is an institute professor and former dean of the School of Engineering at the Massachusetts Institute of Technology. He currently serves as president of the Singapore University of Technology and Design. He has devoted much of his professional career to education that combines engineering and management and to teaching and research in applied and theoretical aspects of large-scale optimization.

Alumni Society Merit Awards

This award honors those alumni who personify the Department of Industrial and Operations Engineering and College of Engineering tradition of excellence and who have achieved significant accomplishments in their professional lives.

1992

Tom Hodgson joined the Ford Motor Company Transmission and Chassis Division as an operations research analyst and later joined the Ford finance staff in the Operations Research Group. He became an assistant professor at the University of Florida in 1970 and a full professor there in 1978. He was the founding director of the Integrated Systems Engineering Institute established at North Carolina State's Industrial Engineering Department in 1984. In 1990, the National Science Foundation invited Hodgson to serve in its Division of Design and Manufacturing Systems, first as program director and then as division director.

Thom Hodgson

1993

Richard Carl Jelinek

Richard Jelinek cofounded the Medicus Corporation in 1970 to apply industrial engineering principals to health care administration. The company spun off a pioneering patient care information system and two publicly held health care information systems companies. He taught industrial engineering and hospital administration at the University of Michigan from 1964 to 1970 and was director of the Systems Engineering Group, Bureau of Hospital Administration, at the University of Michigan, where he developed research programs applying systems and engineering techniques to health services.

1994

Myun Woo Lee served as a professor and chair in the Department of Industrial Engineering at Seoul National University and president of the Korean Institute of Industrial Engineers. He has collaborated with industry and government in both the United States and Korea on numerous research and development projects. He has published more than 90 articles, has supervised more than 100 research contracts, and has registered more than 200 patents.

Myun Woo Lee (shown in center)

1995

C. Robert Kidder joined Borclen, Inc., as chairman and CEO in January 1995. Prior to his appointment with Borden, he was chairman and CEO of Duracell International, Inc. During his time with the company Duracell not only solidified its number one position in the US market for alkaline batteries but also expanded the brand globally to become the world's leading producer and distributor of alkaline batteries.

C. Robert Kidder (shown in center)

1996

J. Dann Engels's companies now operate in India, China, and the Americas and provide products and services in areas including the automotive aftermarket, horticulture, various material handling industrial products, and wholesale and retail battery sales and battery recycling. Along with his wife, Que-Lan, Engels founded the Orchid Foundation to identify and support deserving students in areas of the developing world and provide scholarships at the University of Michigan's College of Engineering. Engels was actively involved in the early stages of MShadow, which matches University of Michigan student athletes with mentors, internships, and postgraduation employment.

J. Dann Engels

1997

Robert J. Vlasic earned his bachelor's degree in industrial and mechanical engineering from the University of Michigan in 1949. He assumed leadership of Vlasic Foods Co. in 1963 when it was still a small local Michigan pickle producer. The business grew to be worth more than $100 million and was ranked the number one pickle maker in the nation. Vlasic served as a director after the company was sold to Campbell Soup Co. He retired as chairman in 1996. He established the endowed position Robert J. Vlasic Dean of Engineering at University of Michigan's College of Engineering.

Robert J. Vlasic (Alumni Society Medal)

1997

John E. Utley Jr. spent 18 years in various capacities with the Timken Company, rising to the position of assistant general manager of sales for the automotive division. Later he served as president and chief operating officer of the Kelsey-Hayes Company. He has been active in several professional organizations, including the Society of Automotive Engineers. He has served on the Wayne State University School of Business Administration's Board of Advisors and as a member of the University of Michigan College of Engineering's National Advisory Committee.

John E. Utley Jr.

1998

Ralph E. Reins

Ralph E. Reins served as the chief executive officer of Qualitor Inc., president and chief executive officer of AP Parts International, Inc., president and chief executive officer of Envirotest Systems Corp., president of Allied Signal Automotive, and president of United Technologies Automotive. He was chairman, chief executive officer, president, and chief operating officer of Mack Truck as well as president and chief executive officer of ITT Automotive. Reins has been lead independent director of Rofin-Sinar Technologies Inc. since November 6, 2013, and its director since September 1996.

1999

Robert M. Brown

Robert M. Brown founded Treystar in 1969 when he began building mobile home parks. In 1978, he and investors began the renovation of the Main Street East complex in Kalamazoo. In the following years, he was involved in financing and starting more than 40 banking, manufacturing, and farming companies. More recently he has focused on promoting entrepreneurism in Michigan and assisting young people in becoming entrepreneurs.

2000

John A. Muckstadt was an active-duty officer in the US Air Force for 12 years, working primarily in the logistics field. He retired from the US Air Force Reserves. Organizations he has consulted for include Accenture (Andersen Consulting), Aspen Technology, Avon, Bell Atlantic, Chicago Pneumatic Tool, Eaton Aeroquip, General Electric, General Foods, General Motors, IBM, Logistics Management Institute, RAND Corporation, SAS Airlines, the US Air Force, the US Navy, Xelus, and Xerox. He is currently the Acheson-Laibe Professor of Engineering in Cornell University's School of Operations Research and Industrial Engineering.

John A. Muckstadt

2001

Roger Kallock has been an integral force in evolving the field of supply chain management. He has led transformations of the longest and most vital supply chains in the world. He served as the deputy under secretary of defense for logistics and materiel readiness for the Clinton administration. Prior to holding that position, he had more than 35 years of logistics experience in the private sector. He has also provided leadership to the supply chain field through over 30 years of involvement with the Council of Logistics Management.

Roger Kallock

2002

Erhan Cinlar is best known for his work on probability models in manufacturing and telecommunications and for his extensive research in Markov processes and martingales. In 1985, he joined the civil engineering and operations research faculty at Princeton University. He served as chair of the department from 1997 until 1999. In 1999, he became the chair of Princeton's newly created Department of Operations Research and Financial Engineering. He was the editor-in-chief of *Mathematics of Operations Research*, the premier research journal in its field, from 1987 to 1992.

Erhan Cinlar

2003

W. Peter Cherry (shown 2nd from right)

W. Peter Cherry spent more than 30 years with Vector Research Incorporated and its successor, the Altarum Institute, before beginning work with Science Applications International. Cherry has contributed to the development and fielding of most of the major systems currently employed by the US Army, ranging from the Patriot Missile System to the Apache helicopter, as well as the command control and intelligence systems currently in use. His later research interests have included peacekeeping operations and the development of transformational organizations and material.

2004

In 2000, Karl G. Bartscht founded Epsilon Health Group to improve the performance of health care providers. He founded Chi Systems, Inc., in 1969. Later he formed Health Care Management, Inc., a Chi affiliate that managed nursing homes and assisted-living facilities. In 2004, he was serving as managing director of Epsilon Health Group, LLC. Bartscht is a pioneer in health care consulting and has planned, designed, and managed health care systems for more than 30 years.

Karl G. Bartscht (shown on right)

2005

Donald C. Graham (Alumni Society Medal)

Donald C. Graham is chairman of the Graham Group, an alliance of independently owned and operated industrial and investment management businesses. In 1960 he started Graham Engineering, which grew from a design services firm into a diversified group of custom manufacturing companies with leading international positions in specialty machinery, plastic packaging, and architectural window systems. This led to the formation of the Graham Group, which has grown to include Graham Capital Company, Graham Partners, Inverness Graham Investments, Striker Partners, and Graham Software Development.

2005

Lawrence D. Burns was serving as GM's vice president of research and development and strategic planning when he completed his 40-year career with the company. In addition to his role as professor of engineering practice in industrial and operations engineering at the University of Michigan, he is director of the Roundtable on Sustainable Mobility with the Earth Institute at Columbia University. His focus at both institutions is energy policy and transportation.

Lawrence D. Burns

A.2. Industrial and Operations Engineering Faculty Members, Active Years, and Areas of Expertise in Alphabetical Order

Last Name	First Name	Year Joined	Year Left	Year Retired	Major Areas of Research
Allen	Wyeth	1955		1963	
Armstrong	Thomas	1977	Active at time of publication		Ergonomics, biomechanics, occupational health, rehabilitation, work design
Aydin	Goker	2003	2009		Operations management
Babich	Volodymyr	2003	2009		Interface of operations, industrial organization, and finance; stochastic models from supply chain management, game theory, finance, and financial engineering
Baker	Kenneth	1970	1974		Operations research, planning. and scheduling
Bartholdi	John	1977	1980		Scheduling, routing, material handling, combinatorics
Baum	Richard	1969	1976		Optimization and control theory, health care systems
Bean	James	1980	2004		Integer programming, genetic algorithms, scheduling, stochastic processes, infinite horizon optimization, equipment replacement, capacity expansion
Berkeley	Richard	1956		1973	Labor incentives, cost control, work measurement
Birge	John	1980	2000		Operations research, mathematical programming, stochastic programming, large-scale programming
Black*	Fred				
Bonder	Seth	1965	1979		Operations research
Boydstun	Louis	1977	1983		Acquisition of psychomotor skills, human performance, work measurement
Bozer	Yavuz	1986	Active at time of publication		Material handling systems; parts flow and control in distribution, logistics, and manufacturing systems; single-story and multistory facility layout; applied operations research in manufacturing and warehousing systems

Bradley	Hugh	1965	1968		Statistical methods and quality control
Byon	Eunshin	2011	Active at time of publication		Predictive modeling and data analytics, fault diagnosis and condition monitoring, reliability analysis, operations and maintenance optimization of wind power systems, modeling and analysis of sustainable energy systems
Chaffin	Don	1965		2007	Ergonomics, occupational causes of musculoskeletal disorders, human simulation software
Chao	Xiuli	2007	Active at time of publication		Queuing, scheduling, financial engineering, inventory control and supply chain management
Chick	Stephen	1995	2003		Stochastic simulation, Bayesian statistics, stochastic optimization, decision analysis, health care delivery, business model innovation in health care
Cohn	Amy	2002	Active at time of publication		Large-scale discrete optimization problems
Daskin	Mark	2010	Active at time of publication		Application of operations research techniques to problems in transportation, supply chain management, facility location modeling and health care
Denton	Brian	2012	Active at time of publication		Computational optimization under uncertainty with applications to health care delivery and medical decision making related to the detection, treatment, and prevention of chronic diseases
Disney	Ralph	1969	1977		Queuing networks and random processes in queuing networks
D'Souza	Clive	2013	Active at time of publication		Ergonomics and human factors addressing human performance, safety and inclusive design concerns in human-in-the-loop systems analysis and design.
Elkerton	Jay	1986	1990		Human computer interaction
Epelman	Marina	1999	Active at time of publication		Operations research, mathematical programming, linear and nonlinear optimization
Fixson	Sebastian	2002	2006		Modularity in design and production, development of technical cost-modeling methods, the study of how technology developments interact with organizational design and management

Flood*	Merrill	1959	1967		
Gage	James	1956		1974	Engineering economics, small plant management
Galliher	Herbert	1963	1982	1982	Operations research in medicine, health, and industrial operations
Guikema	Seth	2015	Active at time of publication		Risk analysis with a focus on data-driven predictive modeling of system reliability in the face of natural hazards, urban and infrastructure sustainability and risk analysis and management, and terrorist risk analysis and the development of risk management decision frameworks
Hancock	Walton	1960	1996	1997	Manufacturing and hospital systems
Henderson	Shane	1996	2002		Discrete-event simulation, stochastic processes, queuing theory, mathematical modeling
Herrin	Gary	1973		2011	Engineering statistics, quality control, experiment design, forecasting/time series analysis
Herzog	Bertram	1965	1972		Computer graphics, computer networks, computer-aided design
Hughes	Richard	2008	Active at time of publication		Occupational biomechanics, orthopedic biomechanics, orthopedic quality improvement
Jelinek	Richard	1965	1971		Health care systems
Jiang	Ruiwei	2015	Active at time of publication		Methodological side: developing data-driven optimization models that incorporate stochastic programming, robust optimization, and statistical analysis Application side: power and water resource system operations with an emphasis on renewable energy integration and system robustness
Jin	Judy	2005	Active at time of publication		Industrial statistics, quality control, and reliability engineering; data fusion and analytics; system modeling, monitoring, diagnosis, and decision making.
Johnson	Clyde	1957		1974	Hospital systems program
Kantowitz	Barry	1999		2012	In-vehicle information and advanced traveler information systems, driver and pilot workload

Kelton	William	1983	1986		Computer simulation and stochastic operations research
Keppo	Jussi	2001	2012		Stochastic control and financial economics
Keyserling	W. Monroe	1984	Active at time of publication		Occupational safety, ergonomics, work measurement, posture analysis
Kieras	David	1984	1993		Cognitive science, cognitive psychology, artificial intelligence, reading comprehension, natural language processing, user interface design
Kimbleton	Stephen	1968	1973		Models and methods of use in the evaluation, selection, and comparison of computer systems; optimal queuing structures; stochastic processes
Kirkwood	Craig	1974	1980		Multiobjective decision analysis
Kochhar	Devinder	1980	1986		Human work with complex automated systems
Lam	Henry	2015	Active at time of publication		Analysis and Monte Carlo methods for large-scale stochastic systems, model uncertainty and simulation output analysis, stochastic optimization.
Lam	Teresa	1988	1994		Applied statistics, stochastic processes, applied probability
Langolf	Gary	1974		1989	Human performance, work measurement, human factors in automotive design
Lavieri	Mariel	2010	Active at time of publication		Applying operations research to health care topics, medical decision making.
Lee	Jon	2011	Active at time of publication		Mathematical optimization. in particular, nonlinear discrete optimization and combinatorial optimization.
Lewis	Mark	1999	2005		Queuing systems, stochastic processes, Markov decision processes, resource allocation and pricing, applied probability
Liker	Jeff	1982	Active at time of publication		Lean manufacturing, lean leadership, operational excellence, design process management

Liu	Yili	1991	Active at time of publication		Cognitive engineering and cognitive ergonomics, human-computer interaction, human factors
Lohmann	Jack	1979	1991		Capital budgeting and engineering economy
Martin	Bernard	1990	Active at time of publication		Human sensory motor performance muscle fatigue, oculo-manual coordination, human vibration, human performance, human factors
Merten	Alan	1970	1981		Computers and information processing systems
Miller	James	1971		2000	Occupational, transportation, and product safety; methods engineering; rehabilitation engineering; expert systems
Murty	Katta	1968		2010	Mathematical programming, network flows, optimization and its application
Nagarajan	Viswanath	2014	Active at time of publication		Design and analysis of algorithms for hard combinatorial optimization problems; approximation algorithms; application areas, include routing, location, scheduling, and data center management
Nair	Vijay	1993	Active at time of publication		Statistics in advanced manufacturing, quality improvement, reliability, process control, design of experiments
Noonan	Francis	1975	1978		Engineering and managerial economics
Olsen	Tava	1994	2001		Stochastic modeling of manufacturing systems, supply chain management, queuing theory, and applied probability
Page	Edward	1951		1968	Manufacturing processes and plant layout
Platzman	Loren	1978	1980		Stochastic processes; control systems
Plumlee	Matthew	2015	Active at time of publication		Understanding the interface between computational and physical systems through designed experimentation and statistical analysis
Pollock	Stephen	1969		2005	Decision analysis, stochastic systems, mathematical modeling, operations research, sequential decision making, reliability modeling

Romeijn	H. Edwin	2008	2014		Optimization theory and applications
Saigal	Romesh	1986	Active at time of publication		Operations research optimization, stochastic analysis and financial engineering
Sarter	Nadine	2004	Active at time of publication		Cognitive ergonomics, design and evaluation of multimodal interfaces, types and levels of automation, adaptive interfaces, human error and error management
Seiford	Lawrence	2000	Active at time of publication		Research in quality engineering focused on process improvement, productivity analysis, and performance measurement utilizing data envelopment analysis for manufacturing and service systems
Sharma	Dushyant	2002	2007		Modeling and algorithm development for discrete optimization
Shen	Siqian	2011	Active at time of publication		Integer programming, stochastic programming and network optimization
Shi	Cong	2012	Active at time of publication		Stochastic modeling and optimization, approximation and online algorithms
Shi	Jianjun	1995	2007		Intelligent manufacturing, dynamic system modeling and control, fusion of statistics and engineering, quality control and improvement, online fault detection, isolation, diagnosis in manufacturing systems
Sibley	Edgar	1966	1972		Database management systems, large-scale information processing systems, generalized graphic systems
Singh	Medini	1990	1993		Real-time scheduling and control, inventory management, flexible manufacturing systems, modeling for operations analysis
Smith	Robert	1980		2012	Operations research, infinite horizon optimization, stochastic processes, queuing theory dynamic programming, intelligent transportation systems
Srinivasan	Mandyam	1985	1992		Queuing networks, performance of local area networks, polling systems, distributed data processing systems, database management systems

Steffy	Wilbert	1955		1976	Computer-generated time standards, inventory control systems, production control systems, economic machine replacement, performance appraisal systems, vendor-evaluated systems, and university-industry combined projects
Teichroew	Daniel	1968	2001		Computer techniques and administrative information systems, development and application of scientific techniques to organizational problems, operation research and management science
Thrall	Robert	1955	1969		Optimization methods
Van Hentenryck	Pascal	2015	Active at time of publication		Constraint programming; mixed nonlinear optimization; computational disaster management, including evacuation planning and power restoration; control and optimization of energy systems, market optimization, logistics, and transportation systems
Van Oyen	Mark	2005	Active at time of publication		Use of worker cross-training, flexible machines and other mechanisms to create agile operations that enhance supply chain performance, applied probability and the control of queuing networks and other discrete-event systems, hospital care delivery improvement, medical decision making
Vines	Quentin	1951	1968		Production planning
Warner	D. Michael	1970	1978		Application of operations research to health care delivery
White	Chelsea	1990	2002		Markov decision processes, logistics and supply chain systems
Wilson	Dean	1961	1968		Computers applied to industrial problems
Wilson	Richard	1956	1983	1984	Operations research, facility planning, production planning
Woo	Anthony	1977	1977		Computer-aided manufacturing, information processing, computational geometry, computer graphics, artificial intelligence
Wu	Chien-Fu Jeff	1993	2004		Quality improvement, experimental design, data mixing, reliability, statistical process control, applications of statistics to industry and engineering

Yano	Candace	1983	1993		Production planning and inventory control, multi-echelon production and distribution systems, vehicle routing, logistics, production scheduling
Zhang	Rachel	1994	2001		Production and inventory control, supply chain management, joint production and financial decisions, optimal bidding strategies in deregulated energy markets

*Additional information is not available.

A.3. Awards and Honors Received by Industrial and Operations Engineering Faculty Active between 2005 and 2015

Name	Awards and Honors
Thomas J. Armstrong	Fellow, American Industrial Hygiene Association The American Institute for Medical and Biological Engineering (Fellow) Human Factors and Ergonomics Society (Fellow)
Yavuz A. Bozer	Award for Technical Innovation in Industrial Engineering from the IIE (1999)
Don B. Chaffin	G. Lawton and Louise G. Johnson Professor of Industrial and Operations Engineering Richard G. Snyder Distinguished University Professor of Industrial and Operations Engineering Fellow, American Institute of Medical and Biological Engineering Lifetime Honorary Fellow, Ergonomics Society of Great Britain Fellow, Human Factors and Ergonomics Society Fellow, American Industrial Hygiene Association Fellow, Society of Automotive Engineers Fellow, American Society of Biomechanics Fellow, International Ergonomics Association American Association for Advancement of Science Arch T. Colwell Award, Outstanding safety research, from Society of Automotive Engineers (1972) Kramer Award, American Academy of Occupational Medicine, outstanding paper written in the field of occupational health (1974) Outstanding Engineering Alumnus Award, University of Toledo (1983) Atwood Outstanding Engineering Service Award, the University of Michigan (1984) Wartenweiler Memorial Lecture, International Society of Biomechanics (1985) Paul M. Fitts Outstanding Educator Award, Human Factors Society (1990) David Baker Outstanding Research Award, IIE (1991) Edward Baier Technical Achievement Award, American Industrial Hygiene Association (1994) Elected to Membership in National Academy of Engineering (1994) Giovanni Borelli Outstanding Scientist Award, American Society of Biomechanics (1999) Award for Outstanding Service, UAW-GM National Joint Committee for Health and Safety (2002) M. Ayoub Award for Outstanding Service to Ergonomics, IIE Society for Work Science (2004) Kettering University, Honorary Doctorate of Engineering (2004) President's Award for Lifetime Achievement, Human Factors and Ergonomic Society (2007) National Engineering Award, American Association of Engineering Societies (2008)
Xiuli Chao	Erlang Prize, Applied Probability Society of INFORMS (1998) David Baker Distinguished Research Award, IIE (2005) Fellow Award, IIE (2014) Jon R. and Beverly S. Holt Award for Excellence in Teaching, College of Engineering, University of Michigan, Ann Arbor (2014)
Amy E. M. Cohn	Alpha Pi Mu IOE Professor of the Year Award (2003, 2004, 2006, 2010, 2012, 2013, and 2015)
Mark Daskin	Presidential Young Investigator Award, NSF (1984) Applied Geography Citation Award, American Association of Geographers, (1985) (with others) Burlington Northern Foundation Faculty Achievement Award (1985) Fulbright Research Award (1989–1990) Listed in Who's Who in America (1996) and Who's Who in American Education Award for Technical Innovation in Industrial Engineering Award, IIE (2001) Fellows Award, INFORMS, (2004) Fred C. Crane Distinguished Service Award, IIE (2005) Fellows Award, IIE, (2006) George E. Kimball Medal, INFORMS (2009) Joint Publishers Book of the Year Award, IIE (2011) David F. Baker Distinguished Research Award, IIE (2014) Lifetime Achievement Award in Location Analysis, INFORMS Section on Location Analysis (2014)

Brian Denton	Daniel H. Wagner Prize for Excellence in Operations Research Practice, INFORMS (2005) Outstanding Publication Award, IIE (2005) CAREER Award, NSF (2009) Secretary of INFORMS (2012–2015)
Marina A. Epelman	Alpha Pi Mu IOE Professor of the Year Award (2001–2002)
Richard Hughes	Excellence in Research Award, American Orthopedic Society for Sports Medicine (2008) John Paul Stapp Award, 53rd Stapp Car Crash Conference (2010) President of American Society of Biomechanics (2013–2014)
Ruiwei Jiang	George Nicholson Student Paper Award, INFORMS (2013) Student Paper Award, Stochastic Programming Society (2013)
Barry H. Kantowitz	American Psychological Society (Charter Fellow) American Psychological Association (Elected Fellow 1974, Society of Engineering Psychologists)
W. Monroe Keyserling	Mellon Fellow, Harvard School of Public Health (1981–1983) Liberty Mutual Insurance Company–International Journal of Industrial Ergonomics Outstanding Paper Award (1994) Invited Speaker, Nordic Institute for Advanced Training and Occupational Health Triennial Conference of the Physiology of Work, Stockholm, Sweden (1994) Keynote Speaker, International Conference on Ergonomics in Occupational Safety and Health, Taipei, Taiwan (1995) Invited speaker and panelist, National Academic of Science Workshop on Work-related Musculoskeletal Disorders (1998) President, Association of University Programs in Occupational Health and Safety (1999–2003) Fellow, American Industrial Hygiene Association (2003) Visiting Scholar, Liberty Mutual Research Institute for Safety (2005) University of Michigan College of Engineering Outstanding Service Award (2008) American Industrial Hygiene Association, Best Ergonomics Paper, American Industrial Hygiene Conference and Exposition, Toronto (2009)
Jon Lee	Computing Society Prize, INFORMS (2010) Elected Fellow of INFORMS (2013)
Mark E. Lewis	Career Award, NSF (2002) Harold R. Johnson Diversity award (2003) Presidential Early Career Award for Scientists and Engineers, NSF (2003) Sloan Foundation's Mentor of the Year (2003)
Jeffrey K. Liker	Shingo Prize for Excellence in Manufacturing Research, *Harvard Business Review*, November/December 1994 (equal author with Rajan Kamath) Shingo Prize for Excellence in Manufacturing Research, *Sloan Management Review*, Spring 1995 (second author with A. Ward, D. Sobek, J. Cristiano) Shingo Prize for Excellence in Manufacturing Research for Liker, J. K., Ettlie, J. E., Campbell, J. C., Roncarti, R., and Tanner, C. (1996) Shingo Prize for Excellence in Manufacturing Research. Liker, J. K. (editor), *Becoming Lean: Inside Stories of U.S. Manufacturers*, 1997. Another eight Shingo Prizes for books published between 2004 and 2012. Best Paper Award, IEEE Transactions on Engineering Management, Summer 1999 (second author with Nazli Wasti) Litrati Club Award for IEEE Transactions on Engineering Management, Summer 1999 (second author with Nazli Wasti) Distinguished Service Award, INFORMS Technology Management Section (2004) Inducted into Association for Manufacturing Excellence Hall of Fame (2012)

Yili Liu	Engineering Students Outstanding Teaching Award (1999) College of Engineering Society of Women Engineers Outstanding Teaching Award (2001) College of Engineering Faculty Education Excellence Award (2001–2002) University of Michigan Arthur F. Thurnau Award/Professorship (2003–present; University of Michigan career-long title) College of Engineering Society of Women Engineers/Society of Minority Jon R. and Beverly S. Holt Award for Teaching Excellence (three-time recipient) Alpha Pi Mu IOE Professor of the Year Award (six-time recipient)
Katta G. Murty	Most Outstanding Faculty Member Award, Alpha Pi Mu (1977–1978) Koopman Prize, Military Applications Society of INFORMS, Outstanding Paper Award (1999) Fellow, INFORMS (2003) Edelman Finalist Award of INFORMS, for work carried out at Hong Kong International Terminals (2004) Fulbright Senior Specialist Award (2006) American Society for Engineering Education Meriam/Wiley Distinguished Author Award (2012)
Vijayan N. Nair	Central Bank of Malaysia Gold Medal (1972) Frank Wilcoxon Prize for Best Applications Paper in Technometrics (1986) Adolphe Quetelet Medal of the International Statistical Institute (1999) Jack Youden Prize for Best Expository Paper in Technometrics (2001) Brumbaugh Award for Best Paper in Industrial Quality Control (2002) Donald A. Darling Professor of Statistics (2002–present) Conference Honoree, Joint Research Conference on Statistics in Industry and Technology (2010) President, International Society for Business & Industrial Statistics (2011–2013) Jack Youden Prize for Best Expository Paper in *Technometrics* (2012) Henry and Byrna David Lecture, National Academies (November 2012) Deming Award and Lecture, Joint Statistical Meetings (August 2013) President, International Statistical Institute (2013–2015)
Stephen M. Pollock	Fellow, Space Technology Laboratory (1960) Fellow, American Association for the Advancement of Science Senior Faculty Fellow, NSF (1975–1976) College of Engineering Stephen S. Attwood Award for Outstanding Contributions to the College (1990) University of Michigan, Society of Fellows (1992–1995) Fellow, Tauber Manufacturing Institute (1999–2002) Herrick Professor of Manufacturing, National Academy of Engineering (2002) INFORMS Fellow (2002) Kimball Medal, INFORMS (2002)
Lawrence M. Seiford	College of Business Administration Foundation Award for Research Excellence, University of Texas (1984) Lilly Endowment Teaching Fellow, University of Massachusetts (1987–1988) Outstanding Advisor Service Award, College of Engineering and Joint Student Engineering Societies, University of Massachusetts (1997) Lilly Faculty Mentor, University of Massachusetts (1999) Doctor Honoris Causa, Universite de la Mediterranee Aix-Marseille (2000)
Robert L Smith	Altarum/Environmental Research Institute of Michigan Russell D. O'Neal Professor of Engineering Outstanding Teacher, Michigan Student Assembly (1989) Fellow, NSF (1970–1971) College of Engineering Research Excellence Award (1999–2000) Distinguished Faculty Achievement Award from the University of Michigan (2003) INFORMS Fellow (2003)
Cong Shi	George E. Nicholson Prize (2009)

Jianjun Shi	CAREER Award, NSF (1996) 1938E Award, College of Engineering, the University of Michigan (1998) University of Michigan Dean's Honor in Teaching Award (1998) Best Paper Award Finalist, North America Manufacturing Research Conference (May 2000) Best Paper Award, American Society of Mechanical Engineers International Mechanical Engineering Congress and Exposition (November 2000) Fellow, Tauber Manufacturing Institute Fellow (2001) Robert M. Caddell Memorial Award, Department of Mechanical Engineering, the University of Michigan (2001) Excellence in Service Award, IIE Transactions (2002) Guest Professor, Beijing Science and Technology University (2002) Excellence in Service Award, IIE Transactions (2003) Excellence in Service Awards, IIE Transactions (2004) Guest Professor, Shanghai Jiaotong University (2004) Guest Professor, Tianjin University (2004) Best Paper Award, Industrial Engineering Research Conference (2006) Guest Professor, Beijing Chemical Engineering University (2006) Endowed Professorship, G. Lawton and Louise G. Johnson Professor of Engineering, the University of Michigan (2007) Fellow, American Society of Mechanical Engineering (2007) Fellow, IIE (2007) Forging Achievement Award, Forging Industry Educational and Research Foundation (2007) Monroe-Brown Foundation Research Excellence Award, College of Engineering, the University of Michigan (2007) National University Technology Network, Distance Education Innovation Team Award (this is a team award for the University of Michigan College of Engineering and GM Technical Education Program (2007) Endowed Professorship, The Carolyn J. Stewart Chair Professor, Georgia Institute of Technology (2008) Fellow, INFORMS (2008) Guest Professor, Chinese Academy of Science (2008) Guest Professor, National Material Science and Safety Center, People's Republic of China (2009) Albert G. Holzman Distinguished Educator Award, IIE (2011) Thanks for Being A Great Teacher, The Center for the Enhancement of Teaching and Learning, Georgia Institute of Technology (2011 and 2012) Elected Member of the International Statistical Institute (2012) Visiting Chair Professor, Peking University (2012) Academician, International Academy for Quality (2013) Best Applied Paper Award, IIE Transactions (2013) Elected Oversea Expert, Chinese Academy of Science (2014)
Mark Van Oyen	Faculty Fellow, ALCOA Manufacturing Systems Faculty (1997) Best Paper Award for IIE Transactions focus issue in Operations Engineering (2000–2001): E. Kim and M.P. Van Oyen, Finite-capacity multi-class production scheduling with setup times. *IIE Transactions* 32, no. 9 (2000): 807–818 Second Place, Best Paper Award (2011), College of Healthcare Operations Management, POMS, for "Patient Streaming as a Mechanism for Improving Responsiveness in Emergency Departments," S. Saghafian, W.J. Hopp, M.P. Van Oyen, J. Desmond, and S. Kronick First Place, Best Paper Award (2011), College of Healthcare Operations Management, POMS, for "Design and Optimization Methods for Elective Hospital Admissions," J.E. Helm and M.P. Van Oyen 2011 Pierskalla Award for Best Paper in healthcare management science, from the Health Applications Section of the INFORMS Society in November 2010 for "Patient Streaming as a Mechanism for Improving Responsiveness in Emergency Departments," S. Saghafian, W.J. Hopp, M.P. Van Oyen, J. Desmond, and S. Kronick First prize: 2012 INFORMS Manufacturing and Service Operations Management (MSOM) Best Student Paper Award to PhD student advisee Soroush Saghafian for "Complexity-Based Triage: A Tool for Improving Patient Safety and Operational Efficiency," with Wallace Hopp, Mark Van Oyen, Jeffrey Desmond, and Steven Kronick First prize: 2012 Doing Good with Good OR Prize, INFORMS Society, awarded to Jonathan Helm, Greggory Schell for the paper "Dynamic Monitoring of Chronic Disease" and related research performed jointly with Mariel Lavieri, Mark Van Oyen, and two members from the Kellogg Eye Institute, Joshua Stein, M.D. and David Musch 2nd Place Best Paper Award (2013), College of Healthcare Operations Management, POMS, for "Operational Planning Models with Service Pathways: Project Portfolio for Phase 1 Trials," by J. Deglise-Hawkinson, B. J. Roessler, and M.P. Van Oyen (2013)

Abbreviations: IEEE, Institute of Electrical and Electronics Engineers; IIE, Institute of Industrial Engineers; INFORMS, Institute for Operations Research and Management Sciences; IOE, industrial and operations engineering; NSF, National Science Foundation; POMS, Production and Operations Management Society.

A.4. Industrial Operations and Engineering PhD Graduates, Topics of Study, and Advisers, 1954–2006

Year	Last Name	First Name	Title	Chair	Cochair
1954	Carson	Robert	Consistency in Rating Method and Speed of Industrial Operations by a Group of Time-Study Men with Similar Training	Gordy	
1958	Harris	Fritz B.	Real Time Control in Discontinuous Production	Allen	
1961	Boston	Ralph E.	A Basis for an Industrial Development Program for British Columbia	Allen	
1962	Golding	Edwin	The Effects of Uncertainty and Monetary Incentives on Investments	Hancock	
1963	Hinomoto	Hirohide	Planning the Expansion of Productive Capacity—A Single-Product System	Allen	
1963	Metzger	Robert W.	The Performance of a Series System of Production Stations Separated by Limited Inventories	Hancock	
1963	Patterson	Richard L.	Some Analytical Methods in the Study of N-Stage Stochastic Service with Applications to the Optimization Problem	Clarke	
1964	Buck	James Roy	Situation Effects on Informal Strategies for Solving Optimum-Seeking Problems	Hancock	
1964	Burkhalter	Barton R.	Emphasis on the 2-Dimensional Pattern-Cutter's Problem	Flood	
1964	Drake	William D.	The Design and Implementation of a Competitive Bidding Strategy	D. Wilson	
1964	Glen	Thaddeus	The Prediction of Work Performance Capabilities of Mentally Handicapped Young Adults	Hancock	
1964	Jelinek	Richard C.	Nursing: The Development of an Activity Model	Johnson	
1965	Cinlar	Erhan	Analysis of Systems of Queues in Parallel	Disney	

1965	Leon-Betancort	Alberto	General-Purpose Optimization Procedures	Flood	
1965	Squire	Dana D.	Product Optimization Considering the Managerial Decision of Price, Marketing and Inventory	Bradley	
1966	Evers	William	A New Stochastic Linear Programming Model	Thrall	
1966	Gustafson	David H.	Comparison of Methodologies for Predicting and Explaining Hospital Length of Stay	Jelinek	
1966	Muckstadt	John A.	Scheduling in Power System Systems	Wilson	
1966	Sargent	Robert G.	A Discrete Linear Feedback Control Theory Inventory Model	Bradley	
1966	White	Charles H.	Sequence Dependent Set-up Times: A Prediction Method and an Associated Technique for Sequencing Production	Wilson	
1967	Chaffin	Don	The Development of Prediction Model for the Metabolic Energy Expended During Arm Activities	Hancock	
1967	De Boor	Carl-Wilhelm R.	The Method of Projections as Applied to the Numerical Solution of Two Point Boundary Value Problems Using Cubic Splines	Bartels	
1967	Moore	J. Michael	Sequencing Problems with Due Date Oriented Objective Functions	Wilson	
1967	Neuhardt	John B.	Application of Mathematical Programming to the Selection of Experimental Factor Arrangements with Resource Constraints	Bradley	
1967	Poock	Gary K.	Prediction of Elemental Motion Performance Using Personnel Selection Tests	Hancock	Jelinek

1968	Haessler	Robert	An Application of Heuristic Programming to a Nonlinear Cutting Stock Problem Occurring in the Paper Industry	Wilson	
1968	Haussmann	R. K. Dietrich	A Queueing Theory Approach to Measuring Quality of Nursing Care: Application in a Burn Unit	Jelinek	
1968	Hess	Carl H.	Effectiveness Of Volley Sequence in Unadjusted Artillery Fire	Bonder	Galliher
1968	Sadosky	Thomas L.	Prediction of Cycle Time for Combined Manual and Decision Tasks	Hancock	
1969	Armstrong	John	Regional Economic Optimization and Effluent Charge Theory	Webber	D. Wilson
1969	Cobian Sela	Jose Miguel	Optimal Operation of Pumped Storage Hydroelectric Systems	Wilson	
1969	Duncan	John	Mathematical and Numerical Methods for Scheduling Cutting Tool Replacement and Application	Galliher	
1969	Freund	Louis	A Model for Measuring the Difficulty of Registered Nurse Assignments		
1969	Ganter	William	Control Policies for the Geometric Quality Model	Evans	Teichroew
1969	Hoag	Laverne	Prediction of Physiological Strain and Performance Under Conditions of High Psychological Stress	Hancock	Chaffin
1969	Solberg	James	A Graph Theoretic Approach to the Study of Network Queues		
1969	Swartzman	Gordon L.	The Statistical Analysis of the Arrival Process of Three Major Types of Hospital Patients	Disney	
1970	Hodgson	Thom	Sequence and Lotsize Policies in a Production-Inventory System	R. Wilson	

1970	Jain	Chittaranjan	A Management Information System to Predict Indirect Labor Staffing Levels Using Queueing Theory	Hancock	
1970	Kilpatrick	Kerry	Model for the Design of Manual Work Stations	Hancock	
1970	Krystynak	Leonard F.	Cost/Effective Regional Control of Rheumatic Fever	Galliher	
1970	Sidney	Jeffrey	One Machine Deterministic Job-Shop Scheduling with Precedence Relations and Deferral Costs	Murty	Thrall
1971	Ash	William	Language and Data Structure for Fact Retrieval	Sibley	
1971	Briggs	Galen P.	Inpatient Admissions Scheduling: Application to a Nursing Service	Hancock	Baker
1971	Kronz	Ronald L.	Effects Of the Response Process in Search Models with False Detections	Bonder	
1971	Sternberg	Stanley R.	Development of Optimal Allocation Strategies in Heterogeneous Lanchester-Type Processes	Bonder	
1971	Su	Shiaw Y.	Optimal Operating Policies for Multiple-Purpose, Multi-Reservoir Systems	Deininger	Pollock
1971	Thomas	Marlin U.	Some Probabilistic Aspects of Performance Times for a Combined Manual and Decision Task	Hancock	
1972	Aswad	Ahmed	A Methodology for Research Allocation to Stochastic Research and Development Activities with a Quadratic Objective	Bonder	
1972	Cherry	William Peter	The Superposition of Two Independent Markov Renewal Processes	Disney	
1972	Di Marco	Attilio E.	The Intermediate-term Security Assessment of a Power Generating System	Wilson	

1972	Merrill	Orin H.	Applications and Extensions of an Algorithm that Computes Fixed Points of Certain Upper Semi-Continuous Point to Set Mappings	Murty	
1972	Moore	Michael L.	A Characterization of the Visibility Process and its Effect on Search Policies	Bonder	
1972	Schutz	Rodney	Cyclic Work-Rest Exercise's Effect on Continuous Hold Endurance	Hancock	
1972	Turner	Roger N.	AMAS— A Method Analysis System Based on Analyst-Oriented Job Descriptions	Hancock	
1973	Garg	Arun	The Development and Validation of a Three Dimensional Hand Force Capability Model	Chaffin	
1973	Hsu	James S.	Systems Analysis of Centralized Reactivation of Exhausted Carbon in Wastewater Treatment	Wilson	Armstrong
1973	Kelkar	Shashikant	Effect of Mercury on Neuromuscular Function and Psychomotor Skills in Occupationally Exposed Workers	Chaffin	
1973	Langolf	Gary	Human Motor Performance in Precise Microscopic Work— Development of Standard Data for Microscopic Assembly Work	Hancock	
1973	Park	Kyung	Computerized Simulation Model of Postures During Manual Materials	Chaffin	
1973	Sayani	Hasan	A Decision Model for Restart and Recovery from Errors in Information Processing Systems	Sibley	
1973	Taube-Netto	Miguel	Two Queues In Tandem Attended by a Single Server	Galliher	
1973	Tobin	Roger	Minimal Complete Matching and Applications	Murty	
1974	Davignon	Gilles R.	Single-Server Queueing System with Feedback	Disney	

1974	Etcheberry	Javier	The Set Representation Problem	Murty	
1974	Martin	James	Computerized Monitoring of Physician-Provided Hospital Based Medical Care	Hancock	
1974	Perry	Ronald F.	A Simulation-Based Planning Model for a Radiology Department	Baum	Warner
1974	Plonka	Francis	A Methodology for Tolerancing, Process Evaluation and Control of Automobile Body Subassembly Designs	Hancock	
1974	Trivedi	Vandankumar	Optimum Allocation of Float Nurses Using Head Nurses' Perceptions	Hancock	Warner
1975	Carlson	David	Computerized Information System Development	Teichroew	
1975	Coffey	Richard J.	Preadmission Testing of Hospitalized Surgical Patients and its Relationship to Length of Stay	Hancock	
1975	Gainer	Edward J., III	Large Scale Convex Quadratic Programming	Murty	Baum
1975	Yamaguchi	Koichi	An Approach to Data Compatibility: A Generalized Data Access Method	Merten	Sibley
1976	Armstrong	Thomas J.	Circulatory and Local Muscle Responses to Static Manual Work	Chaffin	Faulkner
1976	Garg	Arun	A Metabolic Rate Prediction Model for Manual Materials Handling Jobs	Chaffin	
1976	Keefer	Donald L.	A Decision Analysis Approach to Resource Allocation Planning Problems with Multiple Objectives	Kirkwood	Pollock
1976	Navathe	S. B.	Database Restructuring Vol. 1 and Vol. 2	Merton	
1976	Nof Nowmiast	Shimon Y.	Integrated Description and Performance Evaluations of Conveyorized Systems	Wilson	

1977	Brownfield (Gatchell)	Suzanne	Power Boat Operations' Visual Behavior Patterns	Miller	
1978	Fuhs	Paul A.	Hospital Discharge Predictions and their Effect on Admissions Scheduling Systems	Hancock	
1978	Kim	Tae-Moon	Two-Echelon Reorder Point Periodic Inventory Control System with Performance Constraints	Galliher	
1978	Magerlein	David B.	Maximum Average Occupancy and the Resultant Bed Size of Inpatient Hospital Units	Hancock	
1978	Magerlein	James M.	Surgical Scheduling and Admissions Control	Hancock	
1978	Maybee	John D.	Efficient Branch-and-Bound Algorithms for Permutation Problems	Wilson	
1979	Foley	Robert	The M/G/1 Queue with Delayed Feedback	Disney	Platzman
1979	Green	Paul	Rational Ways to Increase Pictographic Symbol Discriminability	Langolf	Weintraub
1979	Hamilton	Richard A.	The Relationship Between the Timeliness of Diagnostic Test Results and Length of Stay Patterns	Hancock	
1979	Kahn	Beverly K.	A Structured Logical Database Design Methodology	Teorey	
1979	Keyserling	W. Monroe	Isometric Strength Testing in Selecting Workers for Strenuous Jobs	Armstrong	Herrin
1979	Lee	Myun W	A Stochastic Model of Muscle Fatigue in Frequent Strenuous Work Cycles	Chaffin	Pollock
1979	Preston	Fred L.	Toward the Design of Coordinated Transit Systems: A Routing and Scheduling Model	Cleveland	Pollock
1979	Purkayastha	Subir	Design Of DBMS-Processable Logical Database Structures	Teorey	

1979	Ruth	Roberta Jean	A Mixed Integer Programming Model for Regional Planning of a Hospital In-Patient Service	R. Wilson	Donabedian
1979	Smith	Phil	Short-Term Memory Scanning Is Related to Memory Span and Mercury Exposure	Langolf	Krantz
1980	Chung	Sung-Jin	Structural Complexity of Adjacency on 0-1 Convex Polytopes	Murty	
1980	Filho	Clovis Perrin	Matching and Edge Covering Algorithms	Murty	
1980	Kang	Maing K.	Procedure For Time-Varying Dynamic Maximal Network Flow	Wilson	
1980	Simon	Burton	Equivalent Markov-Renewal Processes	Disney	Wendel
1980	Wu	Lung C.	Manpower Allocation on Assembly Lines	Wilson	
1981	Al-Idrisi	M. M.	Unconstructed Minimization Algorithms for Functions with Singular or Ill-Conditioned Hessian	Birge	Powers
1981	Johnson	David W.	The Software Development Facility Approach to Improved Software Development	Teichroew	Merten
1981	Miller	George	Sequential Rectifying Inspection with Applicability to Motor Vehicle Emission Certification	Pollock	
1981	Yakin	Mustafa Z.	Multiplier Method Algorithms For Inequality Constrained NLP Problems	Murty	
1982	Gana	Akli	Studies in the Complementarity Problem	Murty	
1982	Lee	Kwan	Biomechanical Modeling of Cart Pushing and Pulling	Chaffin	
1982	Stobbe	Terrence J.	The Development of a Practical Strength Testing Program for Industry	Chaffin	Byers

1983	Anderson	Charles	Biomechanics of the Lumbosacral Joint for Lifting Activities	Chaffin	Herrin
1983	Golhar	Damodar Y.	Sequential Analysis: Non-Stationary Processes and Truncation	Pollock	
1983	Jaridi	Majid	Inspection Error Modeling and Economic Design of Sampling Plans Subject to Inspection Error	Herrin	
1983	Marcellus	Richard L.	Markov Chain Disorder Problems	Pollock	
1983	Thomasma	Timothy	The Triangulation Graph as a Data Structure for Computer-Aided Design	Woo	
1984	Al-Yahya	Yahya	Matching and Covering Algorithms	Murty	
1984	Chung	Min K.	Development of a Statistical Methodology for Improved Analysis of Workplace Injuries	Wu	Herrin
1984	Goldberg	Jeffrey	The Modular Design Problems with Linear Separable Side Constraints: Heuristics and Applications	Bean	
1984	Hopp	Wallace	Non-Homogeneous Markov Decision Processes with Applications to R&D Planning	Bean	Smith
1984	Partovi	Mohammad H.	A Study of Degeneracy in the Simplex Algorithm for Linear Programming and Network Flow Problems	Murty	
1984	Tsui	Louis Y.	Production Scheduling for a Fabrication Assembly System	Wilson	
1984	Umar	Amjad	The Allocation of Data and Programs in Distributed Data Processing Environments	Teorey (CICE)	
1985	Brown	Donald	A Justification for Cross-Entropy Minimization with Applications to Reliability and Risk Assessment	Smith	

1985	Goldberg	Joseph H.	Characteristics of Temporal Processing Within Metal Image Rotation	Langolf	Meyer
1985	Higle	Julia	Deterministic Equivalence in Stochastic Infinite Horizon Problems	Bean	Smith
1985	Lehto	Mark R.	A Structured Methodology for Expert System Development with Application to Safety Ergonomics	Miller	
1985	Ruesch-Evans	Susan M.	Ergonomics in Manual Work-space Design: Current Practices and an Alternative Computer-Assisted Approach	Chaffin	
1985	Vanderveen	David	Parallel Replacement Under Nonstationary Deterministic Demand	Lohmann	
1985	Zabinsky	Zelda B.	Computational Complexity of Adaptive Algorithms in Monte Carlo Optimization	Smith	
1986	Baksh	Shariff	Effectiveness of Capital Budgeting Procedures for Dealing with Risk	Lohmann	
1986	Bloswick	Donald S.	Ladder Climbing: A Dynamic Biomechanical Model and Ergonomic Analysis	Chaffin	
1986	Boyd	Amy	Development of a Minimum Cost Sequential Test to Monitor Injury Incidence on an Operation	Herrin	
1986	Bradley	Joseph	A Participative Ergonomic Control Program in a U.S. Automotive Plant: Evaluation and Implications	Armstrong	Liker
1986	Kotkin	Meyer H.	Operating Policies for Non-Stationary Two-Echelon Inventory Systems for Reparable Items	Yano	
1986	Mittenthal	John	Single Machine Scheduling Subject to Random Breakdowns	Birge	
1986	Radwin	Robert G.	Neuromuscular Effects of Vibrating Hand Tools on Grip Exertions, Tactility, Discomfort and Fatigue	Armstrong	Chaffin

1986	Shin	Sung-Yong	Visibility In the Plane and Its Related Problems	Woo	
1986	Wiker	Steven	Effects of Relative Hand Location Upon Movement Time and Fatigue	Chaffin	Langolf
1987	Abdoo	Yvonne	A Model for Nurse Staffing and the Impact of Inter- Rater Reliability of Patient Classification on Nurse Staffing Requirements	Hancock	
1987	Alden	Jeffrey	Error Bounds for Rolling Horizon Procedures	Smith	Pollock
1987	Chan	Thomas	A New Methodology for Solving the Set Partitioning Problem	Yano	
1987	Dula	Jose	Bounds on the Expectation of Convex Functions	Birge	
1987	Durance	Paul	Application of Logical Design to Incomplete Medical Record Processing	Martin	Hancock
1987	Ganesh	Kothandarama	Serial Replacement Under Evolving Productivity	Lohmann	
1988	Bischak	Diane	Weighted Batch Means for Improved Confidence Intervals for Steady-State Processes	Pollock	Kelton, University of Minnesota
1988	Bolat	Ahmet	Generalized Mixed Model Assembly Line Sequencing Problem	Yano	
1988	Chang	Soo	The Steepest Descent Gravitational Method for Linear Programming	Murty	
1988	Clark	David	Model for the Analysis, Representation, and Synthesis of Hazard Warning Communication	Miller	
1988	Kim	Yeong-Dae	An Iterative Approach for System Setup Problems of Flexible Manufacturing Systems	Yano	
1988	Lee	Hyun	Spatial Decomposition Method and Its Manipulation	Woo	

1988	Lee	Woo-Jong	Tolerancing: Computation on Geometric Uncertainties	Woo	
1988	Lee	Hyo-Seong	Control Policies for Queueing and Production/Inventory Systems	Srinivasan	
1988	Maddox	Marilyn J.	Scheduling a Stochastic Job Shop to Minimize Tardiness Objectives	Birge	
1988	Murray	Joseph	Stochastic Initialization in Steady-State Simulations	Srinivasan	Kelton
1988	Noon	Charles	The Generalized Traveling Salesman Problem	Bean	
1988	Richter Kaplan	Lori-Ann	Resource-constrained Project Scheduling with Preemption of Jobs	Yano	
1988	Ryan	Sarah M.	Degeneracy in Discrete Infinite Horizon Optimization	Bean	Smith
1988	Yu	Chi-Yuang	An Experimental Approach to Work Seat Design: Development of a Chair for Industrial Sewing	Keyserling	
1989	Houshmand	Ali	Discriminant Function Analysis for Autocorrelated Data: Applications in Cutting Tool Monitoring	Herrin	
1989	Lee	Heungsoon	A Methodology for Capacity Planning in Flexible Assembly Systems	Srinivasan	Yano
1989	Lifshitz	Yair	Development and Evaluation of Challenge and Recovery Models of Upper Extremity Cumulative Trauma Disorders Based on Job Attributes Analysis	Armstrong	Keyserling
1989	Woldstad	Jeffrey C.	Electromyographic Analysis of Rapid, Accurate Elbow Movements	Chaffin	Meyer
1990	Al-Sultan	Khaled	Nearest Point Problems: Theory and Algorithms	Murty	
1990	Ben Kheder	Nejib	Economic Lot-Sizing in Just-In-Time Procurement Systems	Yano	

1990	Benson	Peter	A Calculus for Infinite Horizon Optimization	Bean	Smith
1990	Cho	Myeon-Sig	Design and Performance Analysis of Trip-Based Material Handling Systems in Manufacturing	Bozer	
1990	Gan	Jacob	Spherical Algorithms for Setup Orientations of Workpieces with Sculptured Surfaces	Woo	
1990	Hahm	Juho	The Economic Lot Production and Delivery Scheduling Problem	Yano	
1990	Hsiao	Hongwei	Posture Preferences and Postural Behavior During Static, Seated, Visual and Manual Tasks	Keyserling	
1990	Kim	David	Aggregation in Large Scale Markov Chains	Smith	
1990	Park	Yunsun	Average Optimality in Infinite Horizon Optimization	Bean	Smith
1990	Radson	Darrell	Experimental Design in the Presence of an Uncontrollable Variable: Model Characteristics and Design Augmentation in a Front End Alignment Experiment	Herrin	
1990	Rim	Suk-Chul	Circular Layout Problems in Manufacturing Systems	Bozer	
1990	Yang	Kai	New Iterative Methods for Linear Inequalities	Murty	
1991	Arantes	Jose	Resolution of Degeneracy in Generalized Networks and Penalty Methods for Linear Programs	Birge	Murty
1991	Bourland	Karla	Production Planning and Control for the Stochastic Economic Lot Scheduling Problem	Yano	
1991	Brown	Matthew	A Mean-Variance Serial Replacement Decision Model	Lohmann	

1991	Byun	Seong	A Computer Simulation Using a Multivariate Biomechanical Posture Prediction Model for Manual Materials Handling Tasks	Herrin	
1991	Chang	Sung Ho	Statistical Evaluation and Analysis of Form and Profile Errors Based on Discrete Measurement Data	Herrin	
1991	Hughes	Richard	Empirical Evaluation of Optimization-Based Lumbar Muscle Force Prediction Models	Chaffin	
1991	Karabakal	Nejat	The Capital Rationing Replacement Problem	Lohmann	
1991	Palmiter	Susan	Animated Demonstrations for Learning Procedural Tasks	Kieras	Elkerton
1991	Park	Joon Young	Mesh Generation with Quasi-Equilateral Triangulation	Woo	
1991	Saldana	Norka	Design and Evaluation of a Computer System Operated by the Workforce for the Collection of Perceived Musculoskeletal Discomfort: A Tool for Surveillance	Herrin	
1991	Ting	Jame-John	An Object-Oriented Approach to the Development of Integrated Manufacturing Systems	Teichroew	
1991	Ulin	Sheryl	Development of Guidelines for the Use of Powered Hand Tools Using Physical Data	Armstrong	
1992	Baron	Jay	Dimensional Analysis and Process Control of the Body-in-White	Hancock	
1992	Beck	Douglas	Human Factors of Posture Entry into Ergonomics Analysis Systems	Chaffin	
1992	Chen	Lin-lin	Visibility Algorithms for Mold and Die Design	Woo	

1992	Choi	Thomas	Salvation for U.S. Manufacturing? The Role of Value Orientations and Communication Networks in Spreading the Continuous Improvement (CI) Gospel	Liker
1992	Chou	Shuo-Yan	Circular and Parabolic Visibility and their Applications	Woo
1992	Frantz	James P	Effect of Location, Procedural Explicitness, and Presentation Format on User Processing of and Compliance with Product Warnings and Instructions	Miller
1992	Gerth	Richard	Demonstration of a Process Control Methodology Using Multiple Regression and Tolerance Analysis	Hancock
1992	Kerk	Carter	Development and Evaluation of a Static Hand Force Exertion Capability Model Using Strength, Stability and Coefficient of Friction	Chaffin
1992	Liang	Ren	Optimal Sampling Strategies for Surface Roughness Measurement	Woo
1992	Malen	Donald	Engineering for the Customer: Decision Methodology for Preliminary Design	Hancock
1992	Meller	Russell	The Single and Multiple Floor Facility Layout Problem: Applying Simulated Annealing and Mathematical Programming Based Heuristics	Bozer
1992	Noh	Seung Jong	Performance Evaluation of the Distributed Queue Dual Bus Metropolitan Area Network	Srinivasan
1992	Sindi	Ahmed	Acceptance of Information Technology: User Acceptance of Expert Systems	Liker
1993	Adams	Paul	The Effects of Protective Clothing on Worker Performance: A Study of Size and Fabric Weight Effects on Range-of-Motion	Keyserling

1993	Ali	Imtiaz	Laboratory Information Processing Systems	Teichroew	
1993	Erlebacher	Steve	Optimally Allocating Processing Time Variability on a Synchronous Assembly Line	Singh	
1993	Gong	Richard	Validating and Refining the GOMS Model Methodology for Software User Interface Design and Evaluation	Kieras	
1993	Kaufman	David	Direction Choice in Random Walk Algorithms with Application to Global Optimization	Smith	
1993	Krawczyk	Sheila	Psychophysical Determination of Work Design Guidelines for Repetitive Upper Extremity Transfer Tasks over an Eight Hour Workday	Armstrong	Martin
1993	Lee	Byoung-Ki	Variation Stack-Up Analysis Using Monte Carlo Simulation for Manufacturing Process Control and Specification	Hancock	
1993	Park	Hee-Seok	Neurophysiological Analysis of Hand Vibration Effects on Sensorimotor Control	Martin	
1993	Resnick	Marc	An Evaluation of Biomechanics, Kinematics, Psychophysics and Motor Control in the Implementation of Material Handling Devices (MHDs) in Manufacturing	Martin	Chaffin
1993	Rosa	Charles	Modeling Investment Uncertainty in the Costs of Global CO2 Emission Policy	Birge	
1993	Thompson	Deborah	The Perception of Physical Stress as a Measure of Biomechanical Tolerance	Chaffin	
1993	Yeh	Rueyhuei	Optimal Maintenance Policies for Deteriorating Systems	Lam	
1993	Yoon	Jae	Continuous Improvement of Process Control Using Regression Analysis of Observational Data	Hancock	

1994	Abillama	Walid	Optimal Production Policies in Production System with Uncertainties	Singh	Stecke, White
1994	Black-Nembhard	Harriet	A Transient Period Control Methodology for Continuous Mix Manufacturing	Birge	
1994	Hadj-Alouane	Atidel	A Genetic Algorithm for Non-linear Integer Programs	Bean	
1994	Hwang	Juhwen	Production Policy for Systems with Uncertain Capacity and Demand	Singh	
1994	Kawlra	Raj	Development and Application of a Methodology for Minimizing Manufacturing Costs Based on Optimal Tolerance Allocation	Hancock	Patterson
1994	Ledersnaider	David	A Sequential Methodology for Response Service Estimation	Pollock	
1994	Liaw	Ching-Fang	Heuristic Search and Its Transit Applications	White	
1994	Nembhard	David	Heuristic Path Selection in Graphs with Non-Order Preserving Reward Structure	White	
1994	Nussbaum	Maury	Artificial Neural Networks for the Prediction, Classification and Simulation of Lumbar Muscle Recruitment	Chaffin	
1994	Rhoades	Timothy	The Use of Affordance and Categorization Approaches to Evaluate Selected Occupational Movement Behaviors at Auto Hauling Terminals	Miller	Baggett
1994	Takriti	Samer	Stochastic Programs with Dynamically Varying Right-Hand Sides	Birge	
1994	Wunderlich	Karl	Link-Time Prediction for Real-Time Anticipatory Route Guidance in Vehicular Traffic Networks	Smith	
1994	Yi	Tongnyoul	Bipartite Matchings with Specified Values for a 0-1 Linear Function	Murty	

1995	Benhajla	Saida	A Neural Network Decision Model for Managing Product Mix	Birge
1995	Chou	Yu-Li	Accelerating the Solution of Dynamic Programs through State Aggregation	Smith
1995	Cross	William	Approximating Solutions in Infinite Horizon Optimization	Smith
1995	Hammett	Patrick C.	Validating Stamping and Metal Assembly Processes During Automotive Body Development	Hancock
1995	Isken	Mark	Personnel Scheduling Models for Hospital Ancillary Units	Hancock
1995	Keblis	Matthew	Control of Assembly Systems	Duenyas
1995	Kim	Samuel	Framework for Development of Maintenance Policies and Setup Switchovers for CNC Machines	Hancock
1995	Kim	Jonghwa	Transfer Batch Sizing in Trip-based Material Handling Systems	Bozer
1995	Majeske	Karl D.	Interpreting Automobile Warranty Data for Engineering Design and Process Movement	Herrin
1995	Norman	Bryan	The Random Keys Genetic Algorithm for Complex Schedule Problems	Bean
1995	Park	Jung	A Cost-Driven Partitioning Algorithm for Tandem Trip-based Material Handling Systems	Bozer
1995	Wang	Wei-Ching	Detection of Process Change with Non-geometric Failure Time Distribution	Pollock
1995	Wang	Chi-Yueh	The Use of Statistical Quality Improvement Methods in Automotive Body Manufacturing: Three Research Topics	Hancock

1995	Wasti	Nazli	Supplier Involvement in Automotive Component Design: A Study of the U.S. and Japan	Liker	
1996	Alfakih	Abdo Y.	Facets of an Assignment Problem with a 0-1 Side Constraint	Murty	
1996	Chen	Wei-Wang	Managing Variation in Chemical Batch Processes	Herrin	
1996	Donohue	Chris	Stochastic Network Programming and the Dynamic Vehicle Allocation Problem	Birge	Bean
1996	Murray	John	Hortatory Operations the Colloquium: Modeling a Human-Machine System Using Knowledge Engineering Techniques	Liu	
1996	Sang	Shih-Ching Albert	Statistical Modeling of Circular Features and Measurements Obtained Using Coordinate Measuring Machines	Pollock	Ni
1996	Yen	Chih-Kuan	New Strategies for Device Dispatching in Trip-Based Material Handling Systems	Bozer	
1997	Allen	Theodore	Optimal Design of Experiments for Parameter Design and/or Finite Element Analysis	Herrin	
1997	Garcia	Alfredo	Approximating Equilibria for Infinite Horizon Dynamic Games	Smith	
1997	Gerard	Michael J.	Effects of Keyswitch Stiffness, Typing Pace, and Auditory Feedback on Typing Force, Muscle Activity and Subjective Discomfort	Armstrong	Martin
1997	Hsieh	Chung-Chi	An Invariance in the Partial Visibility Due to Mobile Source in Planar Scenes	Woo	
1997	Latko	Wendy	Development and Evaluation of an Observational Method for Quantifying Exposure to Hand Activity and Other Physical Stressors in Manual Work	Armstrong	

1997	Morse	Christopher	Stochastic Equipment Replacement with Budget Restraints	Bean	
1997	Neale	John J.	Control of a Batch Processing Machine	Duenyas	Haessler
1997	Reaume	Daniel J.	Efficient Random Algorithms for Constrained Global and Convex Optimization	Smith	Romejin
1997	Sobek II	Durward	Core Beliefs That Shape Product Development Systems: Explaining Toyota and Chrysler Differences	Liker	Ward
1997	Tsung	Fu-Gee	Run-to-Run Proportional Integral-Derivative Control and Monitoring Schemes	Wu	Shi
1997	Zhang	Xudong	The Development of a Three-Dimensional Dynamic Posture Prediction Model for Seated Operator Motion Simulation	Chaffin	Faraway
1998	Cristiano	John	An Investigation into Best Practice Usage of Quality Function Deployment (QFD) with Proposed Extensions: A US/Japan Comparative Study	Liker	White
1998	Drogosz	John	Pricing and Capacity Decisions Under Demand Uncertainty	Birge	Duenyas
1998	Fang	Ying-Che	Curve Matching by Energy Minimization in Unit Quaternions	Woo	Birge
1998	Hui	Tze-On S.	Application of Multivariate Statistical Methodologies to Body-in-White Assembly Process	Herrin	Scott
1998	Lin	Chih-Jen	Studies in Large-Scale Optimization	Saigal	
1998	Rasch	Steven	World-Class Manufacturing Practices — Do They Work in American Companies?	Liker	
1998	Reed	Matthew	Statistical and Biomechanical Prediction of Automobile Driving Posture	Chaffin	Schneider

1998	Simmons	Julie E.	Maintenance and Replacement Policies for Multi-State Deteriorating Process with Probabilistic Monitoring	Pollock	
1998	Wachs (Okin)	Allise R.	Stochastic Infinite Horizon Optimization with Average Cost Criterion	Smith	
1999	Bander	James L.	Sequential Decision Problems Arising in Commercial Vehicle Operations	White	
1999	Carr	Scott M.	Essays on the Allocation of Scarce Capacity among Multiple Market Segments	Duenyas	Lovejoy
1999	Hashem	Ayman A.	The Effects of Work Climate on Process Innovation in Saudi Arabian Industry	Liker	Birge
1999	Jin	Jionghua	Feature Based Diagnostic System Development Using In-Process Sensing for Stamping Process Control	Shi	
1999	Khan	Ashraf M.	Optimal Sensor Location for Multi-Fixture Assembly System Fault Diagnosis: A Methodology	Woo	Ceglarek
1999	Lin	Zong-Zhi	A Hybrid Genetic/Optimization Algorithm for Piecewise Affine and Convex Markov Decision Processes	Bean	White
1999	Monroe	Kimberly A.	Evaluation of Factors Influencing the Maintenance of Static Non-Seated Work Postures: Assessment of Age, Gender, Strength, Flexibility, Balance, and Subjective Assessment	Keyserling	
1999	Pryor Wellman	Katherine H.	The Trade-off Between Inventory and Transportation Costs in Point-to-Point Truckload Scenarios	White	Duenyas
1999	Steinfeld	Aaron M.	The Benefit to the Deaf of Real-Time Captions in a Mainstream Classroom Environment	Liu	
1999	Supatgiat	Chonawee	Equilibrium Values in Competitive Electricity Markets	Birge	Zhang

1999	Wang	Chi-Tai	Static and Dynamic Facility Layout Problems	Bozer	
1999	Wang	Ming-En	Forward Ray Tracing and Remote Object Locations in Two Dimensions	Woo	
1999	Wu	Yen-chun	Just-in-Time Manufacturing and External Logistics: Evidence from American Parts Suppliers	Liker	
2000	Davydov	Dmitry	Application of Markov Diffusion Processes in Economics and Finance	Linetsky	
2000	Glenn	David W.	Modeling Supplier Coordination in Manufacturing Process Validation	Pollock	
2000	Grasman	Scott	Production Strategies for Random Yield Processes	Birge	Olsen
2000	Inoue	Koichiro	Decision – Theoretic Comparison of Alternative Systems Configurations Using Stochastic Simulation	Chick	
2000	Kiatsupaibul	Seksan	Markov Chain Monte Carlo Methods for Global Optimization	Smith	
2000	Lan	Wei-Min	Dynamic Scheduling of Multi-Product Serial Systems with Setups: Bounds and Heuristics	Olsen	
2000	Lynch-Caris	Terri M.	Ergonomic Justification through Improved Quantitative Output Measures	Herrin	
2000	Oh	Meejung C.	Reengineering Laboratory Information Systems	Teichroew	
2000	Patana-Anake	Prayoon	Two problems in the Optimal Control of Stochastic Manufacturing Systems	Duenyas	
2000	Rong	Qiang	Methodology for Modeling and Diagnosis of Compliant Structure Assemblies	Shi	Ceglarek
2000	Sun	Baocheng	Statistical Process Monitoring for Non-IID Processes	Shi	

2000	Wang	Chi-Kuo	Visibility in Reflective Environments	Woo	
2001	Ahn	Hyun-Soo	Three Essays on Pricing and Dynamic Control	Duenyas	Zhang
2001	Benson	David E.	Load Plan Selection for Package Express Fleets	White	
2001	Choi	Beong	Theory and Algorithms in Semidefinite Programming	Saigal	
2001	Chu	Teresa	A Class of Strictly Semimonotone Matrices in Linear Complementary Theory	Murty	
2001	Garcia-Guzman	Luis M.	Dimensional Management in Metal Assembly Process Validation: A Multivariate Approach	Herrin	Hammet
2001	Hsieh	Ying-Jiun	Three Essays on the Performance Analysis of Closed-Loop Conveyors with Discrete Spaces and Fixed Windows	Bozer	
2001	Lee	Chi-Guhn	Vehicle Routing and Inventory Control for In-Bound Logistics	Bozer	White
2001	Luehmann	April	Factors Affecting Secondary Science Teachers' Appraisal and Adoption of Technology-rich Project-based Learning Environments	Krajcik	Liu
2001	Maillart	Lisa M.	Optimal Observation and Preventive Maintenance Schedules for Partially Observed Multi-state Deterioration Systems with Obvious Failures	Pollock	
2001	Ng	Szu Hui	Sensitivity and Uncertainty Analysis of Complex Simulation Models	Chick	
2001	Torpong	Cheevapra-watdomrong	Monotonicity in Infinite Horizon Optimization	Smith	
2001	Tsai	Chi-Yang	Essays on Optimal Control of Manufacturing Systems with Multiple Product and Demand Classes	Duenyas	

2001	Yen	Joyce W.	A Stochastic Programming Formulation of the Stochastic Crew Scheduling Problem	Birge	
2002	Bailey	Matthew D.	Approximate Solution Techniques for Acyclic Deterministic Dynamic Programming	Alden	Smith
2002	Feyen	Robert G.	Modeling Human Performance using the Queuing Network-Model Human Processor (QN-MHP)	Liu	
2002	Frank	Katia	Two Essays on Optimal Procurement Policies in Supply Chains with Constraints	Zhang	
2002	Fry	Michael J.	Collaborative and Cooperative Agreements in the Supply Chain	Kapuscinski	Olsen
2002	Holloway	Hillary A.	Question Selection for Multi-Attribute Decision Aiding	White	
2002	Huggins	Eric Logan	Supply Chain Management with Overtime and Premium Freight	Olsen	
2002	Lambert	Theodore	Deterministic and Stochastic Systems Optimization	Epelman	Smith
2002	Marshall	Matthew	Development of Models and Procedures for Evaluating Hand Exertion During Manual Work	Armstrong	
2002	Morgan	James M.	High Performance Product Development: A Systems Approach to a Lean Product Development Process	Liker	
2002	Narongwanich	Wichai	Reconfigurable Capacity for Uncertain Markets	Birge	Duenyas
2002	Soorapanth	Sada	Microbial Risk Models Designed to Inform Water Treatment Policy Decisions	Chick	
2002	Thomas	Barrett W.	Anticipatory Route Selection Problems	White	
2002	Yoon	Seunghwan	Optimal Pricing and Admission Control in a Nonstationary Queueing System	Lewis	

2003	Chen	Yong	Integrated Design and Analysis of Product Quality and Tooling Reliability	Shi	
2003	Huang	Qiang	Stream-of-Variation Modeling and Analysis of Multi-Operational Machining Processes	Shi	
2003	Hung	Kuo-Ting	Quality and Financial Implications of Just-in-Time Logistics in Supply Chain Management	Liker	
2003	Kim	Seongmoon	Optimal Vehicle Routing and Scheduling with Real-Time Traffic Information	White	Lewis
2003	Ohlmann	Jeffrey	Theory and Applications of Compressed Annealing	Bean	Henderson
2003	Park	Woojin	Memory-Based Human Motion Simulation	Chaffin	
2003	Ro	Young	Product Architecture & Firm Relations: The Changing Face of Product Development in US Auto	Liker	Fixson
2004	Atlason	Julius	A Simulation-Based Cutting Plane Method for Optimization of Service Systems	Epelman	Henderson
2004	Baumert	Stephen	Stochastic Search Methods for Large-Scale Optimization	Smith	
2004	Grande	Darby	Asset Replacement Considering Environmental and Economic Objectives	Bean	
2004	Hui	Chuck	Multivariate Robust Parameter Design	Herrin	
2004	Karlin	Jennifer	Defining the Lean Logistics Learning Enterprises: Examples from Toyota's North American Supply Chain	Liker	
2004	Kim	Jihyun	In-Process Sensor Fusion and Data Analysis for Forging Process Control and Quality Improvements	Shi	

2004	Kula	Ufuk	Performance Evaluation and Control of Flexible Work Crews and Machines in Manufacturing	Duenyas	
2004	Tsimhoni	Omer	Visual Sampling of In-Vehicle Displays While Driving: Empirical Findings and a Computational Cognitive Model	Liu	Green
2004	Yu	Feng-Tien	Door Allocation Problem at Intermediate Consolidation Terminal of Less-than-Truckload Motor Carriers	Murty	Sharma
2005	Ebersole	Marissa	An Investigation of Exposure Assessment Methods for Selected Physical Demands in Hand-Intensive Work	Armstrong	
2005	Kaufman	David	Dynamic Control of Production Systems with Varying Service Capacity	Lewis	Aydin
2005	Sukhapesna	Sanphet	Generalized Annealing Algorithms for Discrete Optimization Problems	Sharma	
2005	Zhu	Wanshan	Robust Supply Chain Design Mechanisms: Applications to Risk Managements, Coordination, and Multiple-Modular Design	Keppo	Kapuscinski
2005	Kim	Kyung Han	Modeling of Head and Hand Coordination in Unconstrained Three-Dimensional Movements	Martin	
2005	Meng	Xu	Large Agent and Incomplete Markets	Keppo	
2005	Pak	Dohyun	Real Option and Game Theoretic Approach to Telecommunication Network Optimization	Keppo	
2005	Reeves	Kingsley	A Multi-Methodological Study of the Firm Boundary Decision in Automotive Distribution and Logistics: Theory, Practice, and Performance	Liker	

2006	Ghate	Archis	Markov Chains, Game Theory, and Infinite Programming: Three Paradigms for Optimization of Complex Systems
2006	Chaipradubkiat	Pornpen	Integration of Part Quality and tooling Information for Effective Process Control and Maintenance Planning
2006	Wu	Cheng-Hung	Flexible Resource Allocation in Complex Processing Networks with Reliability Considerations
2006	Cheng	Shi-Fen	Game-Theoretic Approaches for Complex Systems Optimization
2006	Kile	Justin	Design of Walk-and-Pick Order Fulfillment Systems
2006	Pomales-Garcia	Cristina	Aesthetic and Performance Aspects of Web-Based Distance Learning Technology
2006	Rider	Kevin	Effects of Ride Motion Perturbation on the Speed and Accuracy of In-Vehicle Reaching Tasks
2006	Schmidt	Kristi	Theoretical and Experimental Investigations in Engineering Aesthetics
2006	Ye	Qing	Essays on Capacity Management Under Uncertainty and Information Asymmetry
2006	Park	Jin-Kyu	Product Architecture as a Strategic Weapon: How Product Architecture Design Changes Affected An Industry's Competitive Structure

A.5. Major Books Published by Industrial and Operations Engineering Faculty, 1970–2005

A. Bruce Clarke and Ralph L. Disney. Solutions Manual: Probability and Random Processes for Engineers and Scientists. New York: J. Wiley, 1970.

This is the first comprehensive textbook for undergraduate courses in introductory probability. It offers a case-study approach with examples from engineering and the social and life sciences. The updated 1985 second edition includes advanced material on stochastic processes. It is used in many junior- and senior-level courses in industrial engineering, mathematics, business, biology, and social science departments.

John R. Griffith, Walton M. Hancock, and Fred C. Munson, eds. *Cost Control in Hospitals*. Ann Arbor: University of Michigan Health Administration Press, 1976.

This pioneering book presents a series of topics and methods needed to reduce hospital costs while improving the quality of patient care.

Katta G. Murty. *Linear and Combinatorial Programming*. New York: John Wiley, 1976.

This is one of the first graduate-level textbooks on this subject and is used by many universities worldwide. A revised edition was published by RE Krieger in 1985.

Leo Greenberg and Don Chaffin. *Workers and Their Tools*. Midland, MI: Pendell Press, 1979.

This was the first reference book to focus on the use of ergonomic principles in the design of hand tools of all kinds.

Katta G. Murty. *Linear Programming*. New York: John Wiley, 1983.

This graduate-level textbook has been adopted by many universities.

Walton Hancock and Paul Walter. *The ASCS: Inpatient Admissions Scheduling and Control System*. Ann Arbor: University of Michigan Health Administration Press, 1983.

This is a technical manual that presents the details of an inpatient admissions model and software. When applied in several Michigan hospitals, the system resulted in significantly improved occupancy rates without many emergency problems.

D. B. Chaffin and G. B. J. Andersson. *Occupational Biomechanics*. New York: John Wiley and Sons, 1984.

This textbook has been used in over 200 universities since it was first published 30 years ago. As such, it has provided the scientific basis for how biomechanical principles, methods, and data can be used to prevent musculoskeletal disorders within ergonomics programs throughout the world. Three subsequent editions included B. J. Martin as a third author.

Katta G. Murty. *Linear Complementarity, Linear and Nonlinear Programming*. Berlin, Germany: Heldermann Verlag, 1988.

This graduate-level textbook has been used by many universities and can now be downloaded from the author's web page at http://www-personal.umich.edu/~murty/.

P. Van Hentenryck. *Constraint Satisfaction in Logic Programming*. Cambridge, MA: MIT Press, 1989.*

This book introduces what is now known as constraint programming. It describes the architecture adopted by all modern constraint programming systems as well as numerous applications of constraint programming in scheduling and resource allocation.

Malcolm Pope, Gunnar B. J. Andersson, John W. Frymoyer, and Don B. Chaffin. *Occupational Low Back Pain: Assessment, Treatment and Prevention.* St. Louis, MO: Mosby Year Book, 1991.

This is the first such book to combine expertise in orthopedics and occupational biomechanics to provide guidance to physicians, physical and occupational therapists, and professional ergonomists to improve the assessment, treatment, and prevention of low back pain in industry.

Stephen Pollock, Michael Rothkopf, and Arnold Barnett, eds. *Operations Research in the Public Sector*. Volume VI of Handbooks in Operations Research and the Management Sciences. Amsterdam: Elsevier, 1994.

This book presents a comprehensive treatment of a variety of public sector applications of operations research methods; it is directed at bridging the gap between theory and practice. Subjects include the military, urban services, crime and justice, health and administration, air and water quality, natural resources management, legislative apportionment, hazardous facility location, voting and paired comparisons, competitive bidding, and theories of measurement.

Romesh Saigal. Linear Programming—A Modern Integrated Analysis. New York: Springer, 1995.

This is the first research monograph that integrates the well-known boundary methods, like the simplex method, and the newer interior point methods. The book also has a chapter devoted to implementations of both methods.

Katta G. Murty. *Network Programming.* Upper Saddle River, NJ: Prentice Hall, 1995.

This is a graduate-level textbook on network flows, a subject that was not available at the time in other books. It was adopted by many universities over many years, and continues to be available from Prentice-Hall. It can also be downloaded from the author's web page at http://www-personal.umich.edu/~murty/.

Wallace Hopp and Mark Spearman. *Factory Physics: Foundations of Manufacturing Management.* Chicago: Waveland Press, 1995.

This book (through subsequent editions) has been used as a text in more than 200 universities and has sold even more copies to industry. The book revolutionized the teaching of operations management by being the first to take a queuing perspective of manufacturing systems. By giving a rigorous description of how these systems behave, this approach provides a systematic framework for improvement. As such, factory physics serves as the underlying science for management practices, including lean and Six Sigma.

Mark S. Daskin. Network and Discrete Location: Models, Algorithms, and Applications. New York: John Wiley and Sons, 1995.*

This textbook discusses key problems and solution methodologies in discrete facility location modeling. It has been cited over 700 times according to Scopus. The text is accompanied by software for solving a number of different classes of location problems. This software has been used by practitioners around the world.

P. Van Hentenryck, L. Michel, and Y. Deville. *Numerica: A Modeling Language for Global Optimization.* Cambridge, MA: MIT Press, 1997.*

This book presents a modeling language and new algorithms for global optimization with applications in chemistry, robotics, and other fields.

Christopher D. Wickens, John D. Lee, Yili Liu, and Sallie E. Gordon Becker. *An Introduction to Human Factors Engineering.* Upper Saddle River, NJ: Pearson Prentice Hall, 1998.

This best-selling ergonomics textbook has been used by many practitioners, educators, and researchers as well as in many universities. The book describes human population strengths and limitations—both physical and mental—and how these should be used to guide the design and evaluation of human-machine-environment systems. A second edition was published in 2004.

Michael Pinedo and Xiuli Chao. Operations Scheduling with Applications to Manufacturing and Services. New York: Irwin/McGraw Hill, 1998.*

This text covers scheduling for operations, both manufacturing and services. Topics discussed include reservations systems, systems design, flexible system scheduling, workforce scheduling, and future scheduling issues, such as web-based systems. The scheduling system developed in this book has been adopted by hundreds of universities around the world.

P. Van Hentenryck. *The OPL Optimization Programming Language.* Cambridge, MA: MIT Press, 1999.*

This book introduces the first modeling language for constraint programming and hybrid optimization. The associated system is a commercial product now sold by IBM.

Xiuli Chao, Masakiyo Miyazawa, and Michael Pinedo. *Queueing Networks: Customers, Signals, and Product Form Solutions.* Chichester, UK: John Wiley & Sons, 1999.*

The objective of this book is to present, in a unified framework, the latest developments in queuing networks with signals. It presents a comprehensive treatment to the field and contains many unique features that never appeared in other books, such as the necessary and sufficient condition for a network process to possess a product-form solution and the general framework for studying discrete time queuing network models.

Henry Wolkowitz, Romesh Saigal, and Lieven Vandenberghe, eds. *Handbook of Semidefinite Programming—Theory, Algorithms and Applications*. New York: Kluwer, 2000.

This handbook offers an advanced and broad overview of the current state of the field. It contains 19 chapters written by the leading experts on the subject. The chapters are organized in three parts: theory, algorithms, and applications and extensions. Semidefinite programming is one of the most exciting and active research areas in optimization. This is the first such book to bring the three areas of work together.

Katta G. Murty. Computational and Algorithmic Linear Algebra and n-Dimensional Geometry. Hackensack, NJ: World-Scientific Press, 2000.

This introductory, self-teaching book has been available on the author's web page for many years and was recently published in print and e-book formats.

James A. Tompkins, John A. White, Yavuz A. Bozer, and J. M. A. Tanchoco. *Facilities Planning.* 3rd ed. New York: John Wiley and Sons, 2002.

This highly regarded, best-selling textbook is used in many industrial engineering/operations research departments for their facilities planning courses, while serving as a principal reference for many practitioners and researchers worldwide. (A 4th edition was published in

2010.) It has been cited over 1,800 times according to Google scholar and has been translated into Chinese, Japanese, Korean, and Spanish. The book presents a wide range of analytic models for facility design, layout planning, and location models.

Katta G. Murtry. Optimization Models for Decision Making, Volume 1. 2003.

This junior-level book that can be downloaded from the author's web page at: http://www-personal.umich.edu/~murty/

Jon Lee. *A First Course in Combinatorial Optimization*. New York: Cambridge University Press, 2004.*

Lee focuses on key mathematical ideas leading to useful models and algorithms in this widely used introductory graduate-level text for students of operations research, mathematics, and computer science. The viewpoint is polyhedral, and Lee also uses submodularity and matroids as unifying ideas.

Jeffrey K. Liker. *The Toyota Way*. New York: McGraw Hill, 2004.

This best-selling business book has sold more than 800,000 copies in 26 languages, and organizations often call it their "company bible." It won a Shingo Prize for Research Excellence and the 2005 Institute of Industrial Engineers Book of the Year Award.

Jeffrey K. Liker and David Meier. *The Toyota Way Fieldbook.* New York: McGraw Hill, 2005.

This accompanying book to *The Toyota Way* is a practical guide that has sold more than 260,000 copies. It is used by many lean practitioners as a primary source for practical advice and won a Shingo Prize for Research Excellence.

P. Van Hentenryck and L. Michel. *Constraint-Based Local Search*. Cambridge, MA: MIT Press, 2005.*

This book presents the idea of constraint-based local search, a declarative language for specifying global search algorithms and the associated Comet system.

**Designates industrial and operations engineering faculty members who wrote their books before joining the department.*

A.6 Past Chairs of the Industrial and Operations Engineering Department

Years	Chair
1955–1963	Wyeth Allen
1963–1968	Walton M. Hancock
1968–1973	Daniel Teichroew
1973–1977	Richard C. Wilson
1977–1981	Don B. Chaffin
1901–1000	Stephen M. Pollock
1990–1992	Chelsea C. White III
1992–1999	John R. Birge
1999–2000	Chelsea C. White III
2000–2010	Lawrence J. Seiford
2010–present	Mark Daskin

CPSIA information can be obtained
at www.ICGtesting.com
Printed in the USA
LVHW011623211121
704018LV00009B/114

9 781607 853671